TRANSPORTATION INFRASTRUCTURE –
ROADS, HIGHWAYS, BRIDGES, AIRPORTS AND MASS TRANSIT

FEDERAL AND STATE ENVIRONMENTAL REVIEWS OF HIGHWAY PROJECTS

COMPARISON OF REQUIREMENTS

TRANSPORTATION INFRASTRUCTURE – ROADS, HIGHWAYS, BRIDGES, AIRPORTS AND MASS TRANSIT

Additional books in this series can be found on Nova's website
under the Series tab.

Additional e-books in this series can be found on Nova's website
under the e-book tab.

TRANSPORTATION INFRASTRUCTURE – ROADS, HIGHWAYS, BRIDGES, AIRPORTS AND MASS TRANSIT

FEDERAL AND STATE ENVIRONMENTAL REVIEWS OF HIGHWAY PROJECTS

COMPARISON OF REQUIREMENTS

MARLENE WILKINSON
EDITOR

nova publishers
New York

Library of Congress Cataloging-in-Publication Data

ISBN: 978-1-63482-390-6

Published by Nova Science Publishers, Inc. † New York

CONTENTS

PREFACE

Three factors—project funding sources and project characteristics, and whether a state allows the adoption of federal review documents generally determine whether a highway project needs a federal environmental review under the National Environmental Policy Act (NEPA) or a state environmental review under state law, or both. This book focuses on environmental reviews of highway projects in states with SEPAs and addresses factors determining whether federal or state environmental reviews are required; how state and federal review requirements compare; and the extent of any duplication in federal and state reviews, including frequency and cost. This book also discusses the role of the environmental review process in federally funded highway projects.

Chapter 1 - Under NEPA, federal agencies evaluate the potential environmental impacts of proposed projects. FHWA has developed a process for NEPA reviews for federal-aid highway projects, such as roads or bridges. According to the Council on Environmental Quality (CEQ) and GAO analysis, 18 states have SEPAs that also require the review of environmental impacts of a variety of actions for highway projects.

The Moving Ahead for Progress in the 21st Century Act (MAP-21) required GAO to examine state environmental reviews for highway projects, including whether they duplicate federal environmental reviews for federal-aid highway projects. This report focuses solely on environmental reviews of highway projects in states with SEPAs and addresses 1) factors determining whether federal or state environmental reviews are required; 2) how state and federal review requirements compare; and 3) the extent of any duplication in federal and state reviews, including frequency and cost. GAO reviewed FHWA and CEQ documents and interviewed officials of these federal agencies; analyzed state laws and regulations; surveyed the 18 states with SEPAs required for highway projects; and interviewed selected state agencies within 9 of those states based on the number of FHWA NEPA reviews underway and other factors.

This report has no recommendations. The U.S. Department of Transportation and CEQ provided technical corrections about federal and state environmental review requirements, which GAO incorporated as appropriate.

Chapter 2 - Under programs administered by the Department of Transportation's (DOT's) Federal Highway Administration (FHWA), certain highway and bridge projects may be eligible for federal funding. Project approval and the receipt of federal funds are conditioned on the project sponsor (e.g., a local public works or state transportation agency) meeting certain standards and complying with federal law. Activities necessary to demonstrate compliance with those requirements may be completed at various stages of project development. Although the names of each stage may vary from state to state, project development generally includes the following: planning, preliminary design and environmental review, final design and rights-of-way acquisition, construction, and facility operation and maintenance.

When there is debate over the time it takes to complete federal highway projects, the environmental review stage has been a primary focus of congressional attention concerning legislative options to speed project delivery. The current process includes activities necessary to demonstrate that all potential project-related impacts to the human, natural, and cultural environment are identified; effects of those impacts are taken into consideration (among other factors such as economic or community benefits) before a final decision is made; the public is included in that decision-making process; and all state, tribal, or federal compliance requirements applicable as a result of the project's environmental impacts are, or will be, met.

Compliance requirements depend on site-specific factors, including the size and scope of the project, and whether and to what degree it may affect resources such as parks, historic sites, water resources, wetlands, or urban communities. For all proposed federal-aid highway projects, however, some level of review will be required under the National Environmental Policy Act of 1969 (NEPA, 42 U.S.C. §4321 et seq.). Broadly, NEPA requires federal agencies to consider the environmental effects of an action before proceeding with it and to involve the public in the decision-making process.

The time it takes to complete the NEPA process is often the focus of debate over project delays attributable to the overall environmental review stage. However, the majority of FHWA-approved projects require limited documentation or analyses under NEPA. Further, when environmental requirements have caused project delays, requirements established under laws other than NEPA have generally been the source. This calls into question the degree to which the NEPA compliance process is a significant source of delay in completing either the environmental review process or overall project delivery. Causes of delay that have been identified are more often tied to local/state and project-specific factors, primarily local/state agency priorities, project funding levels, local opposition to a project, project complexity, or late changes in project scope. Further, approaches that have been found to expedite environmental reviews involve procedures that local and state transportation agencies may implement currently, such as efficient coordination of interagency involvement; early and continued involvement with stakeholders interested in the project; and identifying environmental issues and requirements early in project development.

Bills in the House and Senate (the American Energy and Infrastructure Jobs Act of 2012 (H.R. 7) and Moving Ahead for Progress in the 21st Century (MAP-21; S. 1813)) would reauthorize DOT programs. Both include provisions intended to expedite project delivery by changing elements of the environmental review process, particularly NEPA requirements. This report provides information on existing NEPA and environmental review requirements, particularly requirements that may be subject to change under the House and Senate proposals.

In: Federal and State Environmental Reviews … ISBN: 978-1-63482-390-6
Editor: Marlene Wilkinson

Chapter 1

HIGHWAY PROJECTS: MANY FEDERAL AND STATE ENVIRONMENTAL REVIEW REQUIREMENTS ARE SIMILAR, AND LITTLE DUPLICATION OF EFFORT OCCURS*

United States Government Accountability Office

WHY GAO DID THIS STUDY

Under NEPA, federal agencies evaluate the potential environmental impacts of proposed projects. FHWA has developed a process for NEPA reviews for federal-aid highway projects, such as roads or bridges. According to the Council on Environmental Quality (CEQ) and GAO analysis, 18 states have SEPAs that also require the review of environmental impacts of a variety of actions for highway projects.

The Moving Ahead for Progress in the 21st Century Act (MAP-21) required GAO to examine state environmental reviews for highway projects, including whether they duplicate federal environmental reviews for federal-aid highway projects. This report focuses solely on environmental reviews of highway projects in states with SEPAs and addresses 1) factors determining whether federal or state environmental reviews are required; 2) how state and federal review requirements compare; and 3) the extent of any duplication in

* This is an edited, reformatted and augmented version of the United States Government Accountability Office publication, GAO-15-71, dated November 2014.

federal and state reviews, including frequency and cost. GAO reviewed FHWA and CEQ documents and interviewed officials of these federal agencies; analyzed state laws and regulations; surveyed the 18 states with SEPAs required for highway projects; and interviewed selected state agencies within 9 of those states based on the number of FHWA NEPA reviews underway and other factors.

This report has no recommendations. The U.S. Department of Transportation and CEQ provided technical corrections about federal and state environmental review requirements, which GAO incorporated as appropriate.

What GAO Found

Three factors—project funding sources and project characteristics, and whether a state allows the adoption of federal review documents—generally determine whether a highway project needs a federal environmental review under the National Environmental Policy Act (NEPA) or a state environmental review under state law, or both. Projects without federal highway funding usually do not require a Federal Highway Administration (FHWA) NEPA review, but NEPA reviews of highway projects may still be required to obtain federal permits. Thresholds for environmental review vary under state environmental policy acts (SEPA) and may include project cost or length, whereas NEPA focuses on the potential for significant environmental impacts. Eighteen states have SEPAs required for highway projects, and 17 of these allow for the partial or full adoption of FHWA analyses or documentation to meet state environmental review requirements, according to GAO's survey of these states.

State environmental review requirements are generally similar to the FHWA NEPA process—including consideration of impacts, development and evaluation of project alternatives, mitigation of adverse project impacts, interagency coordination, and public involvement—although differences in specific requirements may affect key environmental decisions. For example, for the consideration of environmental impacts of a proposed highway project, a majority of states responding to GAO's survey indicated that their requirements are similar to FHWA's NEPA requirements overall. However, officials in 7 states GAO surveyed reported that their SEPA requirements related to social and environmental justice impacts are less stringent than FHWA's NEPA requirements. In addition, while state public involvement requirements are generally similar to FHWA's NEPA requirements overall, individual requirements vary, ranging from states that have no requirements to allow public involvement to others that may have more stringent requirements than FHWA's. Officials in 3 states told GAO that in practice they match FHWA's NEPA public involvement requirements for state-only reviews to meet public expectations, even if state law requires less. Further, in the absence of required federal NEPA reviews, certain federal laws related to protection of parklands and historic preservation may not apply to a project, potentially affecting whether a project is determined to have significant impacts and whether those impacts are mitigated.

Officials in 4 of the 18 states in GAO's survey identified instances of potential federal–state duplication in environmental review processes, stemming either from supplemental state requirements or from the lack of alignment between required federal and state review documents. By contrast, 10 of the states in GAO's survey reported that there was no duplication in environmental reviews. Generally, state officials explained that little duplication of effort occurs in state and federal review processes because these reviews are done concurrently by state officials able to address requirements with analyses used for different purposes without replicating effort. Further, 7 of the 10 states reporting no duplication allow for the adoption of a NEPA review to fulfill SEPA requirements. Finally, 4 states pointed to potential duplication or overlap that did not stem from the interaction of state and federal requirements, such as the rework necessary to keep environmental reviews up to date.

ABBREVIATIONS

AASHTO	American Association of State Highway and Transportation Officials
APA	Administrative Proccdure Act
Caltrans	California's DOT
CE	categorical exclusion
CEQ	Council on Environmental Quality
Clean Water Act	Federal Water Pollution Control Act, Act of June 30, 1948
Coast Guard	U.S. Coast Guard
Corps	U.S. Army Corps of Engineers
CRS	Congressional Research Service
DOT	U.S. Department of Transportation

EA	environmental assessment
EIS	Environmental Impact Statement
FHWA	Federal Highway Administration
FHWA's NEPA Toolkit	FHWA's *Environmental Review Toolkit for NEPA and Transportation Decisionmaking*
FONSI	finding of no significant impact
MAP-21	Moving Ahead for Progress in the 21st Century Act
NEPA	National Environmental Policy Act
PAPAI	Project and Program Action Information
ROD	record of decision
SEPA	state environmental policy act
state DOT	state department of transportation

* * *

November 18, 2014

The Honorable Barbara Boxer
Chairman
The Honorable David Vitter
Ranking Member
Committee on Environment and Public Works
United States Senate

The Honorable Bill Shuster
Chairman
The Honorable Nick J. Rahall, II Ranking Member
Committee on Transportation and Infrastructure
House of Representatives

When a state uses federal-aid highway funding for a highway project,[1] the project becomes subject to federal requirements, including a requirement for environmental review established under the National Environmental Policy Act (NEPA).[2] Roughly one-third of the states also have laws that require the review of environmental impacts under a variety of circumstances. Enacted in 1970, NEPA, and the subsequent Council on Environmental Quality (CEQ) implementing regulations, set out an environmental review process that has two principal purposes: (1) to ensure that an agency carefully considers information concerning the potential environmental effects of proposed projects; and (2) to ensure that this information is made available to the public.[3] NEPA and CEQ regulations require federal agencies to analyze the nature and extent of a project's potential environmental effects and, in many cases, document their analyses.[4] NEPA environmental reviews have been identified by critics as a cause of delay for projects because of time-consuming environmental analysis requirements and praised by proponents for, among other things, bringing public participation into government decision making.[5] Congress has also expressed ongoing interest in the cost of and time required for reviews completed under NEPA for federal-aid highway projects, and we have published several reports assessing available information on the cost and time frames of required reviews.[6]

In 2012, the Moving Ahead for Progress in the 21st Century Act (MAP-21)[7] added provisions intended to accelerate highway project completion and reduce costs, while protecting the environment, among other things. Several of these provisions were aimed at expediting project delivery, including the reviews required under NEPA. MAP-21 required GAO to examine state environmental reviews to determine whether they duplicate federal environmental reviews for federal-aid highway projects.[8] To respond to the MAP-21 mandate, we provided key preliminary information to your staff on September 25, 2014. This report transmits our final results related to this review. This report addresses: (1) the factors that determine whether federal or state environmental reviews are required for highway projects, and how the types of federal and state environmental review documents compare; (2) how state environmental review requirements and practices compare with federal requirements for assessing the environmental impact of federal-aid highway projects; and (3) the extent of any duplication in federal and state reviews, including frequency and cost, in states with environmental review requirements for highway projects.

We have defined duplication as occurring when two or more agencies or programs are engaged in the same activities or provide the same services to the same beneficiaries.[9] This report focuses on duplication that might occur between state and federal processes for environmental review of highway projects.

To respond to all three objectives, we reviewed relevant publications, obtained documents from CEQ, the Congressional Research Service (CRS), and the Federal Highway Administration (FHWA), including the FHWA's Environmental Review Toolkit for NEPA and Transportation Decisionmaking, which provides guidance on FHWA's NEPA environmental review process for state department of transportation (state DOT) officials.[10] We also interviewed officials within these agencies and representatives from the American Association of State Highway and Transportation Officials (AASHTO), as well as two academics identified by CEQ as having expertise in federal and state environmental review requirements.

To respond to the first two objectives, we conducted a legal analysis and a survey, which included all 18 states that we and CEQ identified as having state environmental policy acts (SEPA) requiring environmental review of highway projects.[11] Our legal analysis compared key elements of SEPAs and related state regulations to key elements of NEPA and FHWA regulations for NEPA reviews.[12] We identified these elements by reviewing the relevant federal statutes and regulations in consultation with CEQ and FHWA. Specifically, we started with the statutory language of NEPA, which requires agencies to prepare, for major federal actions significantly affecting the quality of the human environment, a detailed statement on, among other things (1) the environmental impact of the proposed action, (2) any adverse environmental effects which cannot be avoided should the proposal be implemented, and (3) alternatives to the proposed action.[13] NEPA, CEQ regulations, and—for federal-aid highway projects—FHWA regulations then specify detailed environmental review processes, which include requirements for interagency coordination, avoiding duplication, and participation by state and local governments as well as the general public, among other things. Based on these statutory and regulatory requirements, and in consultation with CEQ and FHWA, we distilled 12 key elements of NEPA environmental review. Examples of these 12 elements include the type and level of detail required for environmental reviews, requirements to improve coordination and promote efficiency, and public participation. Details on all 12 elements and findings from our legal comparison of the NEPA and SEPA processes based on these elements are contained in appendix I, and we refer to these findings throughout this report as our legal analysis. To compare state environmental review requirements and practices in more detail for highway projects subject to FHWA requirements, we surveyed state DOTs in each of the 18 states that have SEPAs required for highway projects on the degree of similarity between state requirements and federal requirements identified in FHWA's Toolkit.[14] For more information about how we conducted and used the survey, see appendix II. In addition to these analyses, we interviewed state DOT officials and officials with state resource agencies (e.g., state departments of natural resources) in 9 of the 18 SEPA states we surveyed, which we selected based on a range of factors, including the number of FHWA NEPA reviews underway. Our findings for these nine states are not generalizable to the other states with SEPAs but provide examples of varying state requirements and practices.

To respond to the third objective, we included in our survey of the 18 state DOTs with SEPAs questions related to duplication and included questions about duplication in our interviews of state officials and FHWA division officials in the nine states we selected for additional interviews. In the context of environmental review requirements, such duplication could occur if, for example, two separate—but similar—analyses were carried out to satisfy federal and state requirements, but would not occur if the same analysis could be used to satisfy both state and federal documentary (i.e., procedural) requirements. We inquired about duplication of review efforts within and among highway projects, as well as duplication that may occur across time within a project. We also asked FHWA and state officials for any data or analyses regarding the cost of any potential duplication (and how such cost might be measured) and the frequency of any potential duplication. Finally, in our survey we asked state officials about efforts to improve the efficiency and timeliness of the environmental review process, as well as any potential benefits from such efforts. See appendix II for additional details on our objectives, scope, and methodology.

We conducted this performance audit from May 2013 to November 2014 in accordance with generally accepted government auditing standards. Those standards require that we plan and perform the audit to obtain sufficient, appropriate evidence to provide a reasonable basis for our findings and conclusions based on our audit objectives. We believe that the evidence obtained provides a reasonable basis for our findings and conclusions based on our audit objectives.

BACKGROUND

Under NEPA, federal agencies generally are required to evaluate the potential environmental effects of proposed federal actions such as permitting, funding, approving, or carrying out of federal-aid highway projects. These evaluations allow decision makers to fully consider significant environmental impacts, develop and evaluate project alternatives, and consider mitigating adverse impacts before they decide whether to approve a proposed project. The NEPA process also provides a forum for ensuring that all applicable federal and state resource protection laws and requirements are addressed during the project development process, such as applicable provisions of the federal Clean Water Act[15] and Endangered Species Act.[16] NEPA's implementing regulations require coordination as a means to avoid duplication and increase efficiency. For example, they require all federal agencies to "cooperate with state and local agencies to the fullest extent possible to reduce duplication."[17] In addition, the environmental review of projects can include, among other things, seeking input, and in some cases approvals, from the public as well as federal and state agencies responsible for natural resources, environmental protection, and historic preservation, and obtaining approval of the environmental evaluation by the lead federal agency prior to making a decision, which in some cases results in the issuance of a record of decision (ROD).

FHWA has statutory and regulatory requirements for conducting NEPA reviews, which include additional highway-specific requirements and put additional emphasis on an interagency decision-making process.[18] For instance, FHWA regulations require that each project alternative considered in the review connect logical end points (e.g., involve the development of a highway that links with existing roads) and be of sufficient length to address environmental matters broadly.[19] FHWA regulations and policy also include provisions to coordinate with other federal agencies with jurisdiction, expertise, or interest in the project, as well as non-federal actors, including state agencies and project sponsors.[20] In addition, FHWA's statutory authority includes a mechanism to limit the time frames available to challenge decisions in federal court as part of the judicial review process,[21] as well as other provisions, such as permission to fund assistance to specific groups to improve interagency coordination.[22] Figure 1 illustrates six principles of FHWA NEPA review from its Toolkit.

According to CEQ and FHWA, establishing the purpose and need for the federal action takes into account the project sponsor's purpose and need for a project and is essential in establishing a baseline for the development of the range of reasonable project alternatives required in environmental reviews. It also assists with the identification and eventual selection of a preferred alternative. For FHWA NEPA reviews for federal- aid highway projects, the purpose and need for a proposed project might take into account the status of the project (e.g., the project's history, funding, and schedules, as well as information on agencies involved); a discussion of highway or bridge capacity, the project's relationship to the larger transportation system, and traffic demand; as well as social demands or economic development.[23]

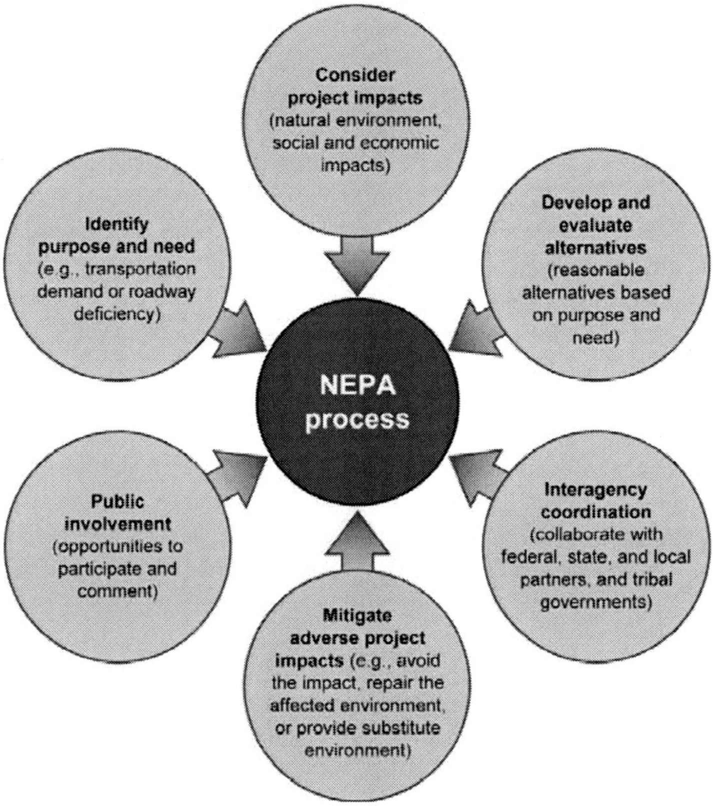

Source: GAO.| GAO-15-71.

Figure 1. The Federal Highway Administration's Principles of the National Environmental Policy Act (NEPA) Process in Transportation Decision Making.

In addition to the procedural requirements of NEPA, the decision to fund a federal-aid highway project also must comply with substantive environmental and natural resource protection laws, including applicable state laws.

- Federal environmental laws commonly applied to a proposed highway project such as the Clean Water Act (Section 404); the Endangered Species Act of 1973 (principally Section 7); Section 106 of the National Historic Preservation Act;[24] and Section 4(f) of the Department of Transportation Act (protecting publicly owned parks, recreation areas, wildlife and waterfowl refuges, and public or private historic sites).[25] Section 4(f) applies to DOT projects and Section 106 applies to projects with federal funding and some projects that require federal permits. Other federal natural resource protection laws—FHWA has identified over 40—can apply to federal and state or local projects, depending on characteristics of the project. For example, the Wild and Scenic Rivers Act protects designated and potential wild, scenic, and recreational rivers.[26]

- State environmental laws, such as laws related to the growth-inducing effects of agency actions may also apply to a proposed highway project. States may also have laws that protect certain resources, such as those designating protected or endangered species or those protecting tribal or other cultural resources.

For federal-aid highway projects, reviews are typically conducted by state DOT officials—or other project sponsors—who carry out analyses and coordinate with FHWA. FHWA generally serves as the federal lead agency for the NEPA process and approves the environmental impact documentation.[27] Under NEPA, the level of review required depends on the potential significance of the environmental effects of the project.[28]

- *Projects that have the potential for a significant effect:* An Environmental Impact Statement (EIS) must be prepared for a project that has the potential for a significant effect on the environment. Both CEQ and FHWA regulations identify EIS requirements related to public involvement (such as public participation in scoping, response to substantive public comment on a draft EIS, and public hearings when appropriate); interagency participation; consideration of project impacts (such as impacts on water quality or wildlife habitat); development and evaluation of alternatives; and the mitigation of adverse project impacts. Projects requiring an EIS are likely to be complex and expensive, as we have noted in prior work.[29] The median time to complete a highway project EIS was over 7 years in 2013, according to FHWA data,[30] and the EIS itself may cost several million dollars, as reported by FHWA.[31]
- *Projects that may or may not have a significant effect:* When project effects are uncertain, project sponsors (such as state DOTs or local ighway departments) must prepare an environmental assessment (EA) to determine whether the project may have a potentially significant impact on the human environment. An EA briefly provides evidence and analysis sufficient to determine whether to prepare an EIS or a

finding of no significant impact (FONSI). A FONSI presents the reasons why the agency has concluded that no significant environmental impacts will occur if the project is implemented. FHWA regulations governing EAs require early scoping and coordination activities, as well as making the EA publicly available. EA review activities have been estimated to take from 14 to 41 months, according to reports from FHWA and AASHTO.

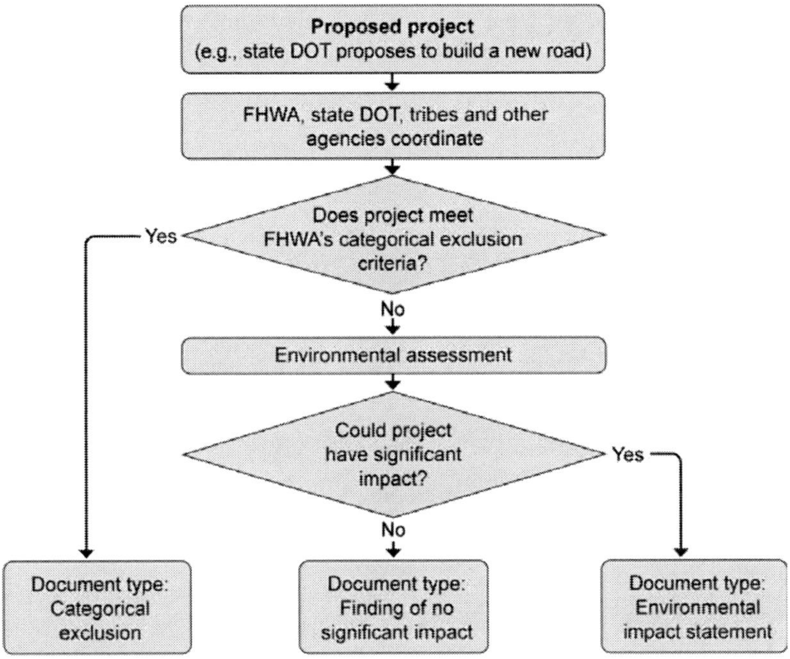

Sources: GAO and FHWA. | GAO-15-71.

Figure 2. National Environmental Policy Act Decision Making Process—Types of Documentation Required for Federal Highway Administration (FHWA) Projects.

- *Projects that normally will not have a significant effect:* These projects, by their very nature (e.g., the project fits within a category of activities that the agency has determined normally do not have the potential for significant environmental impacts), require limited review under NEPA to ensure—by considering any extraordinary circumstances—that a proposed project does not raise the potential for significant effects. Agencies promulgate categorical exclusions (CE) for such projects in their NEPA implementing procedures.[32] The subsequent use of a categorical exclusion for a proposed project, as described in CEQ guidance, can reduce paperwork and delay by speeding the review process. CEs typically take much less time to prepare than EAs or EISs. Environmental review activities for categorically excluded projects have been estimated to take an average of 6 to 8 months to complete, according to FHWA, and could take as long as an average of 22 months to complete, according to a report prepared for AASHTO. Federal-aid highway projects that are generally processed as CEs include resurfacing, constructing bicycle lanes, installing noise barriers, and landscaping projects.

Figure 2 illustrates FHWA's NEPA decision-making process, including the three main types of highway project environmental review documentation.

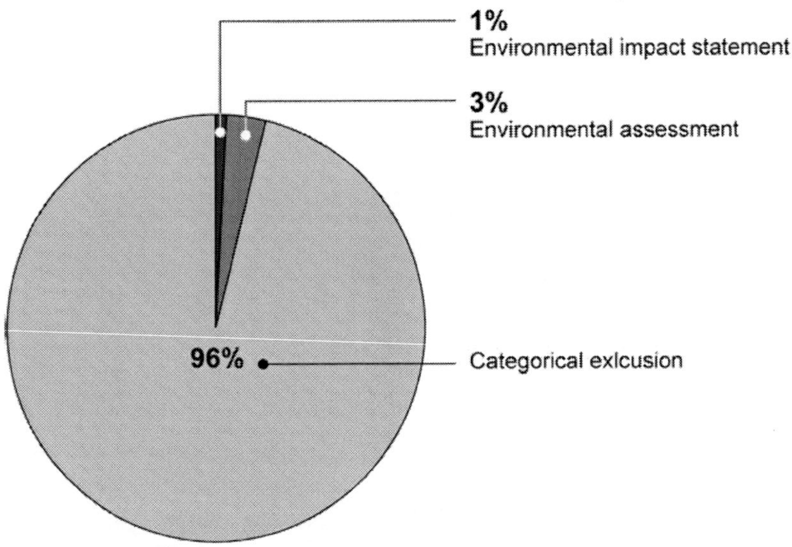

GAO analysis of 2009 FHWA data. | GAO-15-71.

Note: According to the Federal Highway Administration, 2009 data are the most recent available, and the proportion of projects in each category has remained constant based on estimates made with fiscal data.

Figure 3. Federal-Aid Highway Projects by National Environmental Policy Act Review Type.

For federal-aid highway projects, FHWA has found that the vast majority of projects (96 percent in 2009) qualify for environmental review as CEs, and only 1 percent require EIS reviews, as shown in figure 3 below. However, EIS projects, because of their high costs, account for a greater share of federal-aid funds than their numbers might suggest.

According to CEQ and our analysis, 18 states have adopted SEPAs that require environmental reviews for highway projects (see figure 4). The majority of SEPAs were modeled on NEPA and require state and sometimes local public agencies to assess the impacts of projects (or other actions) affecting the quality of the environment within the state.[33] Among other things, SEPAs may expand requirements for environmental reviews to projects (e.g., state, local, or private projects) that are not required to have such reviews under federal law.

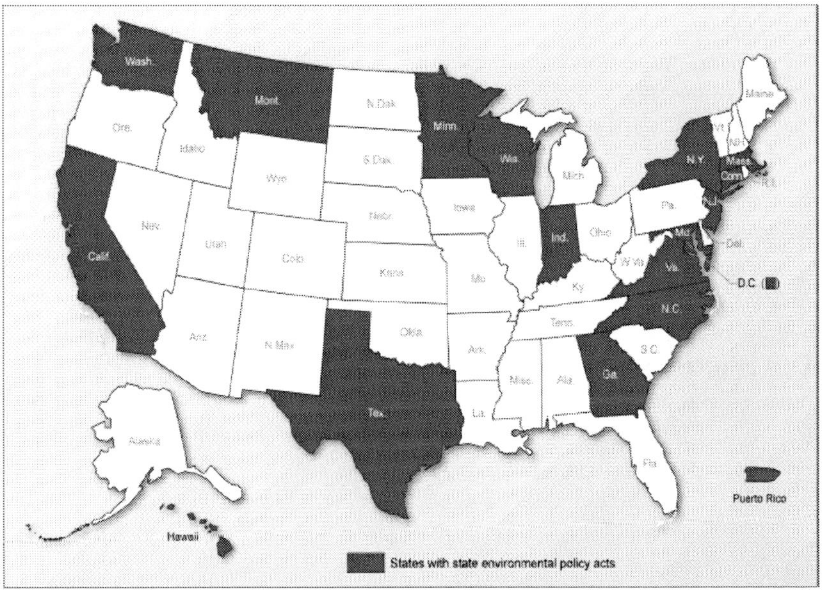

Sources: GAO analysis and Map Resources. | GAO-15-71.

Figure 4. States with State Environmental Policy Acts Required for Highway Projects, as of 2014.

Three Factors Generally Affect the Types of Environmental Reviews Required for Highway Projects, and State Environmental Review Document Types Largely Mirror NEPA

Three Factors Generally Affect Whether Federal or State Environmental Reviews Are Required for Highway Projects

Three factors—project funding sources, project characteristics, and rules allowing state adoption of federal review documents—generally determine whether a highway project needs a federal environmental review or a state environmental review, or both.[34]

Funding Sources

Federal-aid highway projects are generally subject to environmental review under NEPA and the environmental provisions of title 23, with FHWA serving as the federal lead agency for the review. By contrast, when a project is funded solely through state or local funds, it rarely requires an FHWA NEPA review, although action by another federal agency may require an environmental review.[35]

To some extent, states can influence whether or not FHWA NEPA reviews are required by their selection of funding for a particular project. By using—or avoiding—federal-aid highway funds, states can determine whether their projects are subject to a FHWA NEPA review and even whether certain other federal environmental laws apply.[36] For example, officials in some states told us that federal-aid highway funding was sometimes requested for a project requiring permits from other federal agencies specifically so that FHWA would serve as the federal lead agency for the NEPA review, rather than have a federal permitting agency (e.g., the U.S. Army Corps of Engineers (Corps)) serve in that capacity.

Officials in two states cited their positive working relationships with FHWA, or the lack of resources at other federal agencies, to explain such decisions. As another example, California officials explained that not having federal funding or not needing a federal permit—either of which would trigger NEPA—requires them to comply with more burdensome requirements of the Endangered Species Act. They therefore would have an incentive to include federal-aid highway funds to ensure NEPA would apply.[37]

Some state officials, however, identified reasons why they may seek not to use federal-aid highway funding for certain projects. For example, state officials in Massachusetts, Minnesota, and North Carolina told us that they chose not to use federal-aid highway funding on some projects mainly to avoid certain federal review requirements, notably Section 4(f) or Section 106, which can increase project costs or require additional time for reviews to meet federal requirements.[38] For similar reasons, officials in three other states (California, Maryland, and New York) focused their federal funding on certain—often large—projects, resulting in fewer FHWA NEPA reviews overall. Some state officials told us that as a practical matter, there may not always be a choice to not use federal funding, even if they might otherwise choose to do so, because limited state funding may not cover the full cost for some large projects. Although only using state funds may preclude the need to meet certain federal requirements, other federal environmental requirements, such as compliance with the Clean Water Act, largely remain for state-funded projects.

Project Characteristics

SEPAs vary with respect to which project characteristics trigger an environmental review as well as what type of review is required and the scope and extent of that review. These characteristics can include thresholds related to project costs, project length, and expected service impacts such as the volume of traffic affected, among other things. These requirements contrast with NEPA, which generally focuses on the potential for significant environmental impacts.[39] For example, in Virginia, reviews are required for state projects costing $500,000 or more. Other dollar-value thresholds are built into the SEPA requirements of the District of Columbia, Georgia, and New Jersey.

In Massachusetts, environmental reviews are required for the construction of new roadways 2 or more miles in length. Likewise, Minnesota requires an EIS-type review for new road projects four or more lanes in width and 2 or more miles in length.[40] Finally, in certain cases, states require particular environmental reviews for expected service impacts. For example, EA-type reviews are required for the construction of new roads over 1 mile in length that will function as collector roadways in Minnesota.

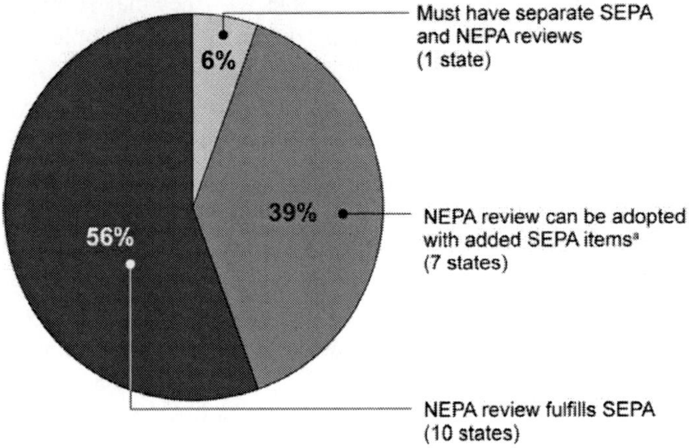

Source: GAO analysis. | GAO 15-71.

[a] California indicated both that a NEPA review can be adopted with added SEPA items and that "it varies." Based on the state's added narrative comments, we included California with the states where a NEPA review can be adopted with added SEPA items, although the adoption process can vary depending on the project sponsor.

Figure 5. The Number of Surveyed States with State Environmental Policy Acts (SEPA) Reporting Extent of Adoption of Federal National Environmental Policy Act (NEPA) Reviews to Meet SEPA Requirements, as Indicated by State Departments of Transportation Survey Responses.

State Adoption of Federal Reviews

Almost all states with SEPAs responding to our survey (17 of 18) allow for the partial or full adoption of analyses or documentation produced in conducting federal reviews to meet state requirements for highway projects, while Massachusetts requires that a separate state review be completed. Figure 5 illustrates the number of states allowing for full adoption, in which the NEPA review fulfills the SEPA requirements; those allowing for the adoption of the federal review with additional state analyses or documentation (i.e., partial adoption); and those requiring that state requirements be met separately from the FHWA NEPA review.[41] Several factors may affect whether a NEPA document can be accepted as a SEPA document in those states that allow for full or partial adoption. For example, state DOT officials reported that their ability to use a NEPA review to meet SEPA requirements varies based on the project sponsor, the type of review, and whether a similar type of environmental review document is required to satisfy federal and state environmental review requirements. Finally, several states carry out and adopt federal analyses or documentation for state projects even in the absence of federal funding or permitting. Maryland DOT officials told us that in practice they adopt federal FHWA NEPA reviews for almost all highway projects—even those with only state funding—given staff familiarity with the federal requirements and public expectations, as well as the potential for funding sources to change between the preliminary engineering and construction phases of a project. Washington DOT officials also mentioned that using and adopting NEPA could be advantageous because they felt more certain about how FHWA's NEPA requirements would be interpreted by courts should a legal challenge to the environmental review process be subsequently filed.

While the majority of states with SEPAs allow for the full adoption of federal NEPA reviews, some do not (see figure 5). When separate federal and state reviews are required, the processes are often carried out concurrently, with joint planning processes, research and studies, and public hearings, as well as the use of blended documents. Both Montana and New York reported having integrated processes for state and federal reviews, for example. Likewise, officials with California's DOT (Caltrans) stated that Caltrans has a long-standing practice of combining NEPA and the state's SEPA processes to make the delivery of transportation projects more efficient. For example, Caltrans's guidance for a blended or joint NEPA/SEPA EIS/EIS-type review describes use of a special chapter in the joint document to address required California-specific mitigation information. Finally, Hawaii's environmental review statute requires coordination of state and federal reviews when both apply. The extent to which state SEPAs, like NEPA, require coordination is discussed in the next section of this report.

When a project sponsor requests that a state review be adopted to satisfy federal requirements, FHWA conducts a legal sufficiency review to ensure that the analysis and documentation satisfy FHWA's NEPA requirements, according to an official with FHWA's Office of Project Development and Environmental Review. This legal review may happen when, for example, federal funding is added to an ongoing state project or when project requirements change, and a federal permit that was not originally required must now be obtained. State officials in several states we spoke with pointed to the potential for duplicative effort when ongoing state projects are subsequently "federalized" in this manner.

According to the FHWA official we spoke with, a case-by-case assessment is necessary because each situation is different when projects have been federalized. To avoid the risk of having to start the federal environmental impact assessment late in the project development process, FHWA encourages project sponsors to follow FHWA's NEPA process from the beginning of project development. For this reason, officials in five states—Hawaii, Maryland, North Carolina, Washington, and Wisconsin—told us they preferred to review projects under FHWA NEPA—even in the absence of federal funding.

State Review Document Types Largely Mirror NEPA Review Document Types

A majority of states we surveyed reported having state review document types that are similar to those used for NEPA reviews (see figure 6). Most states (16 of 18) reported requiring documents that are similar to a NEPA EIS or a NEPA EA, although fewer states had similar documents for FONSI/EAs or CEs, and the level of analysis required varies by state, as discussed in more detail below.

While most states reported having an EA-type document, some state officials told us that their documentation requirements for that process can differ from the documentation requirements for an FHWA EA. FHWA's EAs are to include brief discussions of the need for the proposed project, project alternatives, the environmental impacts of the proposed project and alternatives, and a listing of the agencies and persons consulted. Officials in six of nine states where we conducted interviews told us they use a checklist for their EA-type reviews. For example, Washington DOT's checklist asks for narrative responses to 12 pages of questions on topics like water, plants, and historic and cultural preservation.

Instructions direct the project sponsor—often the state DOT—to answer the questions briefly with the best description possible and to say "does not apply" if that is the case. As a result, these reviews may look more like a documented federal CE than a federal EA. Further, in some instances, these checklist reviews can involve less analysis than a federal EA. A Washington DOT official told us that the checklist allows impacts to not be assessed if they are deemed "not applicable." By comparison, a federal EA requires more detailed documentation of findings. In Massachusetts, the state DOT prepares a 22-page checklist called an environmental notification form for projects meeting set criteria.

This document has attached plans and is publicly circulated. Depending on whether the project exceeds certain thresholds, the document may require responses to questions, according to Massachusetts DOT officials. By contrast,

New York requires more specific analysis as its regulations require the state DOT to prepare a negative declaration that can be supported by a NEPA FONSI.[42]

This document, a determination of no significant effect, must identify all the relevant areas of environmental concern and show why the project impact, if any, is not significant.

MANY STATE AND FEDERAL REQUIREMENTS ARE SIMILAR, BUT DIFFERENCES COULD AFFECT STATE ENVIRONMENTAL REVIEWS OR OUTCOMES

State Requirements Are Generally Similar to FHWA's NEPA Requirements

To compare state environmental review requirements to FHWA's NEPA requirements for federal-aid highway projects, we reviewed relevant state statutes and regulations for each of the18 states that we identified as having a SEPA required for highway projects and compared those statutes and regulations with NEPA and FHWA regulations.[43] In addition, we surveyed state DOTs in each of those 18 states about the degree of similarity between state requirements and federal requirements identified in FHWA's NEPA Toolkit for five of FHWA's six NEPA principles: assessment of project impacts, development and evaluation of project alternatives, mitigation of adverse project impacts, interagency coordination, and public involvement.[44] Seventeen of the 18 surveyed states provided responses to this section of our survey.[45]

We used our legal analysis and interviews with state officials to supplement and confirm survey responses, as well as to provide illustrative examples of how state requirements or practices compare with NEPA and FHWA requirements.[46]

To identify which states have requirements that were "generally similar" to FHWA's NEPA requirements overall, we determined which states in our survey reported having environmental review requirements that were similar or somewhat similar to 42 individual requirements for FHWA NEPA reviews under the five NEPA principles.[47]

Based on our legal analysis and survey responses, for each of the five NEPA principles the majority of states have requirements that are generally similar to FHWA's NEPA requirements overall. More specifically, as shown in figure 7, for each of the five areas, survey results indicated that requirements in 10 or more of the 17 state SEPAs were generally similar to FHWA's NEPA requirements overall. Some similarity between state and federal requirements is to be expected since a majority of SEPAs were modeled on NEPA, as we discussed above.

Further, some states may employ more stringent review processes in practice than state statutes and regulations require in order to satisfy public expectations or for other reasons. Based on our legal analysis and survey responses, we found that a number of states have SEPA requirements that are generally less stringent, however. In fact, survey results suggest that the divergence between state and FHWA's NEPA requirements may be greatest for the NEPA principle of alternatives analysis, where 7 (of 17) states had requirements they characterized as generally less stringent. Overall, no state reported having requirements more stringent than FHWA NEPA for more than 4 (of 42) individual requirements within the five NEPA principles included in our analysis. We discuss each of these NEPA principles—and the related requirements identified in the Toolkit—in more detail below.

In some cases, the divergence between state and federal requirements is more pronounced depending on the type of review being conducted. According to survey responses, state requirements for documentation and analyses are more likely to mirror FHWA's NEPA requirements for higher level, more complex SEPA reviews (e.g., EIS-type reviews) than for less complex reviews.

This difference reflects, in part, the differences among various document types mentioned above, including the use of checklists for EA-type assessments in many states. Potential environmental impacts for the vast majority of projects at both the state and federal level are evaluated using CE or EA requirements, as only a small proportion of projects typically requires an EIS review, according to FHWA and state officials we interviewed. Where state and federal laws diverge, there is the potential for meaningful differences in how significant impacts are assessed or mitigated, which project alternative is selected, the level of interagency coordination, and opportunities for the public to affect or challenge decisions, among other things.

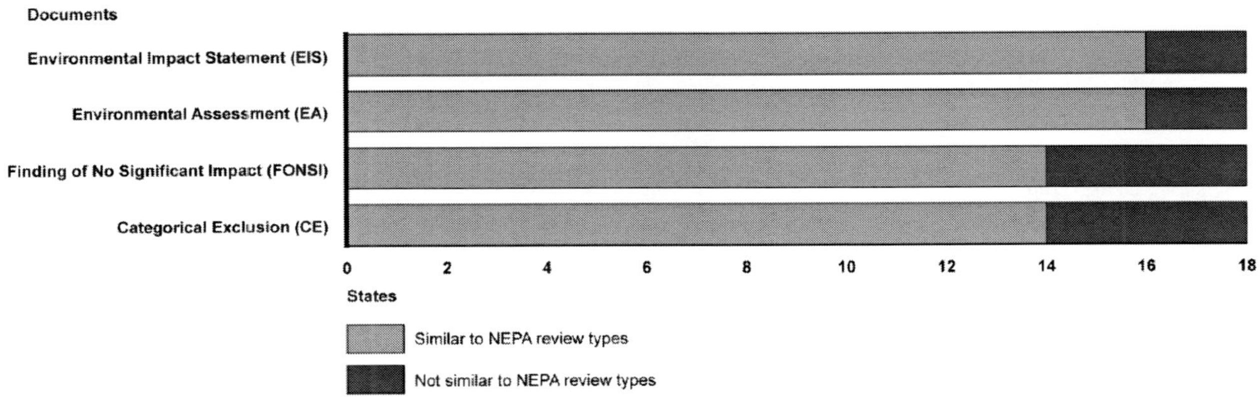

Source: GAO. | GAO-15-17.

Note: States may use different names for their documentation or review type.

Figure 6. Number of States with Environmental Review Documents Used for Their Respective State Environmental Policy Acts (SEPA) That Are Similar to Those Used for the National Environmental Policy Act, as Indicated by State Department of Transportation Survey Responses.

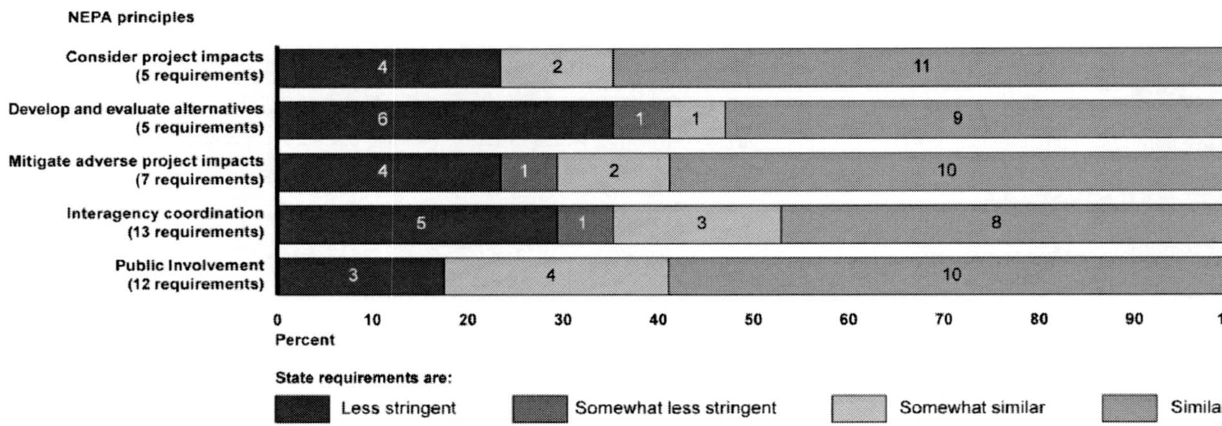

Source: GAO analysis of survey results. | GAO-15-17.

Note: We surveyed state DOTs in states with state environmental policy acts (SEPA) required for highway projects comparing state requirements and federal requirements identified in FHWA's *Environmental Review Toolkit for NEPA and Transportation Decisionmaking.*

Figure 7. Comparison of State Environmental Review Requirements with Federal Highway Administration (FHWA) National Environmental Policy Act (NEPA) Requirements, as Indicated by State Departments of Transportation (state DOT) Survey Responses.

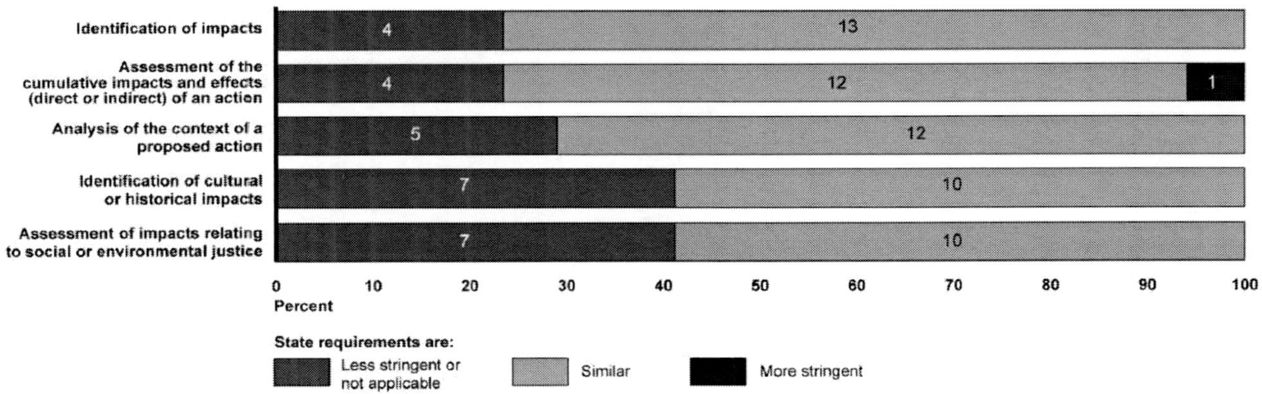

Source: GAO analysis of survey results. | GAO-15-17.

Figure 8. Comparison of State Requirements with Federal Highway Administration (FHWA) National Environmental Policy Act (NEPA) Requirements for the Consideration of Project Impacts, as Indicated by State Departments of Transportation (state DOT) Survey Responses.

Most States Require Some Evaluation of Impacts, Alternatives, and Mitigation, but Requirements Differ, and Comparable Parkland and Historic Preservation Protections Are Generally Absent

Most state requirements include some consideration of impacts, development and evaluation of alternatives, and mitigation for the analyses that inform environmental review documents, but the degree of similarity to FHWA's NEPA requirements varies, and states generally lack protections comparable to federal parkland and historic preservation protections.

Consideration of Project Impacts

For each of the 5 individual requirements related to the consideration of project impacts, from 10 to 13 (of 17) states reported in our survey that their requirements for analyzing the environmental impacts of a proposed highway project are similar to FHWA's NEPA requirements, although other states reported having less stringent requirements.[48] Figure 8 illustrates this variation among states for each of the five requirements associated with the consideration of project impacts.

Impacts analysis provides decision makers with the information necessary to determine whether a proposed action will produce significant adverse impacts in certain identified areas, including impacts that are short-term, long-term, and cumulative.[49] In our survey, 10 (of 17) states reported similarities with FHWA's NEPA requirements to consider cultural or social impacts and to assess impacts relating to social and economic justice, but officials we interviewed in 3 states identified consideration of impacts related to social and environmental justice as a key difference between state requirements and FHWA's NEPA requirements. (See figure 8.) These requirements assess potential effects on certain minority or low-income populations in FHWA NEPA reviews, among other things.[50] For example, Caltrans officials explained that usually they are not required to look at social or economic impacts unless these impacts are triggered by a physical impact, which differs somewhat from federal requirements to address these impacts separately.[51] Similarly, requirements in Wisconsin and Washington for assessing social justice impacts are less stringent than federal requirements, according to state DOT officials, although in practice DOT procedures for analysis in both states align with what is required by FHWA and NEPA.

Variations in state substantive environmental statutes and regulations (e.g., those managing growth or protecting certain species) can affect the determination of significant impacts. For example, North Carolina officials described substantive state environmental statutes and regulations that are more stringent than federal protections, such as higher permit standards to protect trout waters and stricter navigation requirements for public use of waterways. These statutes and regulations affect what must be considered during impacts analysis. Such requirements, while not part of SEPAs themselves, can affect the evaluation of project impacts when they are included in the SEPA or NEPA review processes. According to our legal analysis, Georgia regulations are less stringent because although they require consideration of the cumulative impacts of other proposed government actions, they do not address the actions of nongovernmental entities, for example.

Development and Evaluation of Project Alternatives

For each of the five individual requirements related to the development and evaluation of project alternatives, from 9 to 11 (of 17) states reported in our survey that their respective requirements are similar or more stringent, although notable differences between state and FHWA's NEPA requirements affect both the assessment of alternatives and the selection of the preferred alternative in some states. Figure 9 illustrates this variation among states for each of the five requirements associated with the development and evaluation of project alternatives.

FHWA regulations require consideration and objective evaluation of all reasonable project alternatives to avoid any indication of bias toward a particular alternative, including the "no action" alternative.[52] In our survey, 10 (of 17) states reported similarities with FHWA's NEPA requirements for the identification of a range of reasonable alternatives.[53] (See figure 9.)

In our legal analysis, we found that the District of Columbia requires that EIS-type reviews include a discussion of reasonable alternatives to the proposed governmental action, including the option to take no action.[54] However, about a third of the SEPA states reported less stringent standards for identifying alternatives in our survey, and we found in our legal analysis that just over half of states with SEPAs (10 of 18) require assessment of the "no action" alternative. In Minnesota, for example, only the preferred alternative is assessed for EA-type reviews, which, as state DOT officials agreed, is less stringent than FHWA's NEPA requirements as it may preclude the consideration of some alternatives, although the state does require analysis of the no-action alternative for EIS-type reviews.[55] In addition, we found that some states do not have detailed requirements for the consideration of alternatives, and two states— Virginia and Indiana—do not include requirements for the identification of alternatives in their laws or regulations.[56]

In addition to the identification of alternatives, we also surveyed states about their requirements for selecting the preferred alternative. Seven (of 17) states we surveyed reported that their requirements for alternatives selection are less stringent than FHWA's NEPA regulations, which require that the alternative selection process lead to decisions that are in the best overall public interest.[57] Even when state requirements are similar to FHWA's NEPA requirements for the selection of alternatives generally, there can be some variation in requirements.[58] For example, officials in North Carolina told us that state requirements for alternative selection are similar to FHWA's NEPA requirements, and that in practice state officials select the least environmentally damaging practical alternative as the preferred alternative, although the selection process differs from the FHWA NEPA process.

Mitigation of Adverse Project Impacts

The extent to which states have requirements to consider mitigation of environmental impacts as part of their environmental review process that are similar to FHWA's NEPA requirements often varies by the type of review, and 4 of the 17 states reported in our survey that they do not have mitigation requirements similar to FHWA's NEPA requirements for any level of review. Figure 10 illustrates this variation among states for each of the seven requirements associated with considering mitigation of environmental impacts.

Mitigation is intended to avoid or minimize any possible adverse environmental effects of an action where practicable. FHWA's NEPA regulations require consideration of mitigation regardless of whether the impacts of a proposed project are found to be significant, and the regulations require implementation of mitigation measures if doing so represents a reasonable public expenditure.[59] Based on our legal analysis of state requirements for mitigation, we found that some state requirements are similar to FHWA's NEPA requirements, but may include different terms (potentially altering the thresholds for the consideration of and implementation of mitigation) or use different processes for the evaluation or identification of mitigation.[60] For example, Hawaii's laws require mitigation to be considered as part of its alternatives analysis, but this consideration is only required when mitigation measures are proposed.[61] Moreover, an official in FHWA's Hawaii Division Office told us that environmental mitigation measures, once considered, are not binding under Hawaii's SEPA.

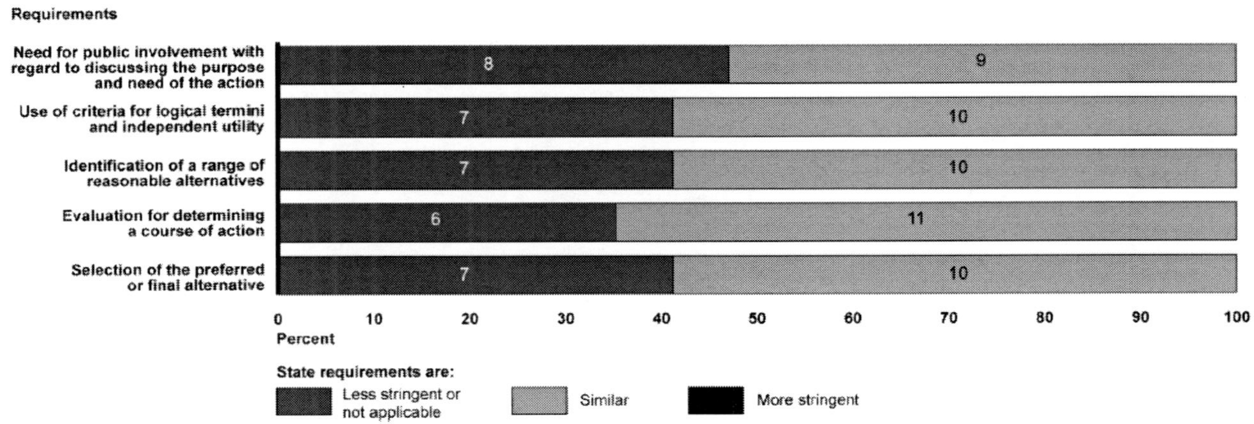

Source: GAO analysis of survey results. | GAO-15-17.

Note: In the course of project development, project sponsors determine end points to be rational or "logical termini" in the context of the purpose of the transportation improvement and the environmental impacts of a proposed project. FHWA requires that project sponsors use criteria to make this determination.

Figure 9. Comparison of State Requirements with Federal Highway Administration (FHWA) National Environmental Policy Act (NEPA) Requirements for the Development and Evaluation of Alternatives, as Indicated by State Departments of Transportation (state DOT) Survey Responses.

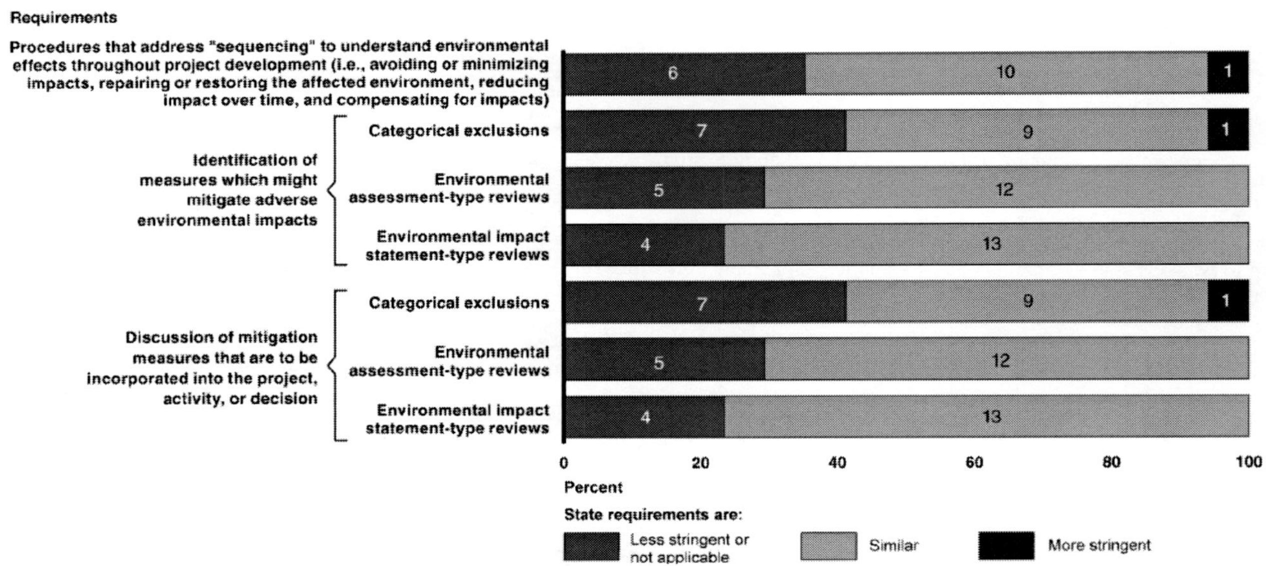

Source: GAO analysis of survey results. | GAO-15-17.

Note: One state reported two instances of having more stringent mitigation requirements than FHWA NEPA and another state reported one instance of having more stringent mitigation requirements.

Figure 10. Comparison of State Requirements with Federal Highway Administration (FHWA) National Environmental Policy Act (NEPA) Requirements for the Mitigation of Adverse Project Impacts, as Indicated by State Departments of Transportation (state DOT) Survey Responses.

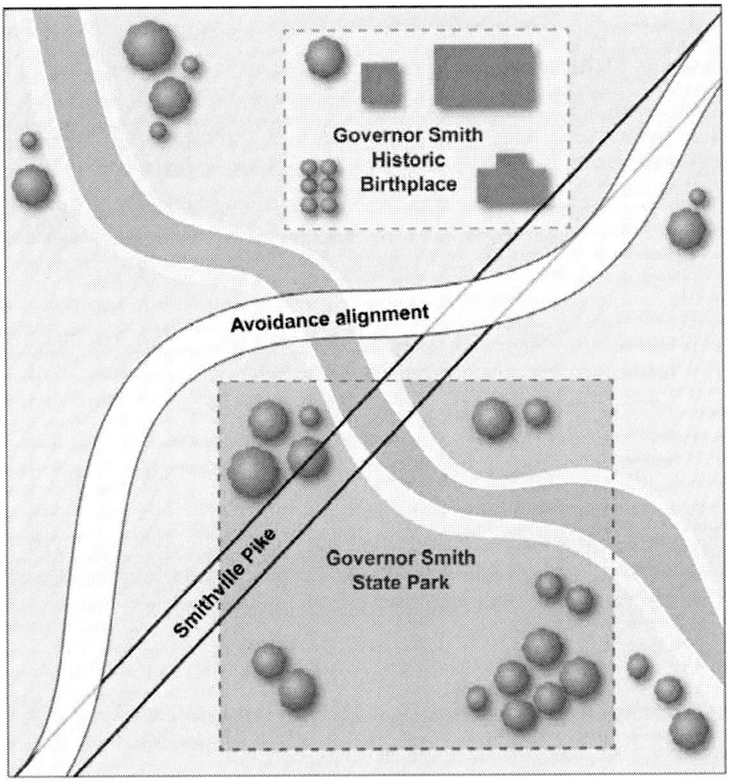

Source: GAO. | GAO- 15-17

Figure 11. Hypothetical Example of an Avoidance Alternative for Parklands.

Parkland and Historic Preservation Protections

As previously discussed, two federal substantive resource protection laws generally do not apply in the absence of FHWA or other federal involvement: Section 4(f) protecting parklands and Section 106 for historic preservation. While other federal environmental laws, such as the Clean Water Act and the Endangered Species Act, apply to all federal, state, or local projects, Section 4(f) and Section 106 are generally not required for highway projects without federal involvement.[62] In 6 (of 9) states where we conducted in-depth interviews, officials reported that their state laws do not address or do not provide protection comparable to these federal laws. For example, in North Carolina, a state without comparable parkland protections, officials told us that for projects not subject to Section 4(f), the state does not require an analysis to provide alternatives that will specifically avoid affecting park or recreational land. (Figure 11 illustrates what such an avoidance alternative might involve under FHWA's NEPA requirements.) In addition, officials said that North Carolina has a much less stringent historic-preservation law than the federal Section 106. Officials in other states also indicated that state requirements were not as stringent as Section 106, particularly with regard to the consideration of project impacts on resources that were eligible for protection, but were not yet on the federal registry of protected resources. Conversely, specific laws in some states provide protection that may be more stringent than what is required under federal law for certain resources. For example, Hawaii has an historic protection requirement for certain cultural resources that goes beyond the requirements in Section 106. Having less stringent or no state requirements for parkland protection and historic preservation may affect whether a project is determined to have significant impacts and therefore whether, for those states that require mitigation, those impacts are considered and mitigated. In addition, less stringent state requirements can provide an incentive to avoid using federal funds for a project, according to state officials in 3 of the 9 states where we conducted in- depth interviews.

State Interagency Coordination Requirements Are Generally Similar to FHWA's NEPA Requirements, with Exceptions

For each of the 13 individual requirements related to interagency coordination of environmental reviews, from 8 to 12 (of 17) states reported in our survey that their respective requirements are similar or more stringent. Figure 12 illustrates the variation among states for each of the 13 requirements associated with interagency coordination for environmental reviews.

Under NEPA, FHWA, as the lead federal agency, is required to coordinate the timing and scope of environmental reviews to develop consensus among a wide range of stakeholders with diverse interests. These coordination requirements are intended to make the review process more efficient, eliminate duplication, and reduce delays, by including tribes, businesses, transportation or environmental interest groups, resource and regulatory agencies, neighborhoods, and affected populations, among others, in the environmental review process. As part of state requirements for interagency coordination, some states encourage cooperation and consultation with federal or state agencies, among others, but state requirements vary in how similar they are to FHWA's NEPA requirements. (See figure 12.) Specifically, state requirements regarding the coordination of outreach to specific populations—such as underserved or minority groups—are often less stringent than FHWA's NEPA requirements.

In fact, 9 (of 17) state DOTs we surveyed reported that for EAs, their state requirements for coordinating outreach to specific populations are less stringent than FHWA's NEPA requirements, with 8 (of 17) state DOTs reporting less stringent requirements for EIS-type reviews. In states reporting less stringent requirements, there may be less involvement from minority or underserved populations, affecting how and whether potentially affected populations are involved in the review of proposed projects.

In our legal analysis, we found that a few states had requirements that corresponded to the "single-process review" included in FHWA's NEPA requirements to promote efficiency and avoid delays by including insofar as practical, the completion of all environmental permits, approvals, reviews, or studies as part of the NEPA process, including those by other federal agencies such as permits from the Corps under the Clean Water Act. Some state officials we interviewed told us that their state encouraged, but did not require, cooperation and consultation with federal or state agencies during the environmental review process. For example, while not requiring the completion of all permits or reviews, Washington's SEPA requires that environmental reviews include a list of all licenses (e.g., permits) that will be needed for the project.[63]

State DOT officials in several states told us that in practice they employ systems that may go beyond minimum state requirements to coordinate with federal and state agencies, including FHWA, for the review of highway projects. These officials described various state efforts to develop consensus among stakeholders, ranging from regularly scheduled meetings to the use of state clearinghouses to ensure timely stakeholder receipt of documentation for comment. For example, Wisconsin DOT officials have developed an agreement with the state's Department of Natural Resources to meet and coordinate on impacts analyses prior to issuing a draft EIS-type document. Also, in North Carolina, regulations empower the state agency responsible for compliance with the state's SEPA to seek information from federal as well as local and special units of government. According to North Carolina officials, sometimes a formal interagency process is used, even in the absence of a federal NEPA review, because it allows for better coordination.

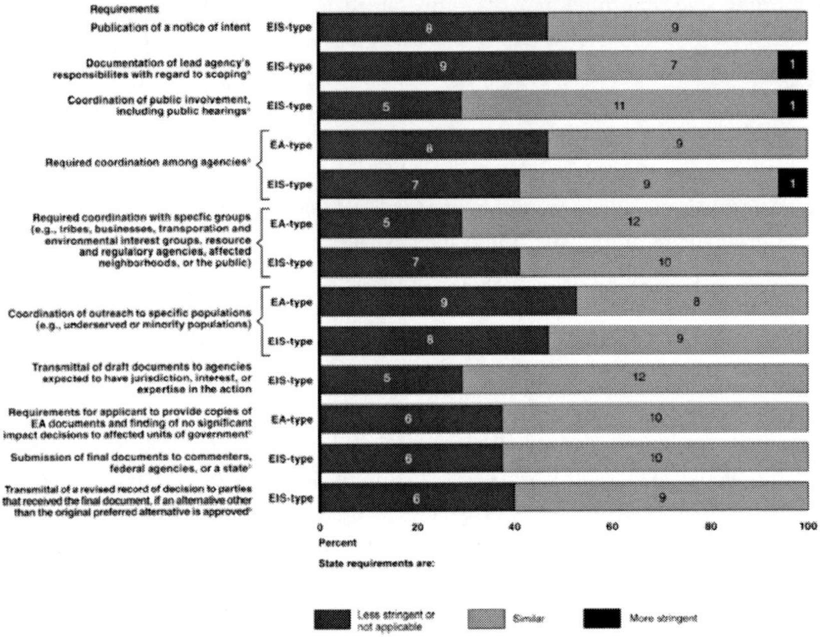

Source: GAO analysis of survey results. | GAO-15-17.

[a] Three different states account for the three instances in which states reported more stringent requirements.

[b] Some survey response totals amount to fewer than 17 because some state DOTs did not respond to all of the questions.

Figure 12. Comparison of State Requirements with Federal Highway Administration (FHWA) National Environmental Policy Act (NEPA) Requirements for Interagency Coordination for Environmental Assessments (EA) and Environmental Impact Statements (EIS), as Indicated by State Departments of Transportation (state DOT) Survey Responses.

States' Public Involvement Requirements Are Generally Similar to FHWA's NEPA Requirements, Though a Number of States Have Less Stringent Requirements when Conducting EAs

State environmental review requirements for public involvement are generally similar to FHWA's NEPA requirements overall, although public involvement requirements for EA-type reviews varied. Figure 13 illustrates the variation among states for each of the 5 EA-type review requirements and 7 EIS-type review requirements associated with public involvement.

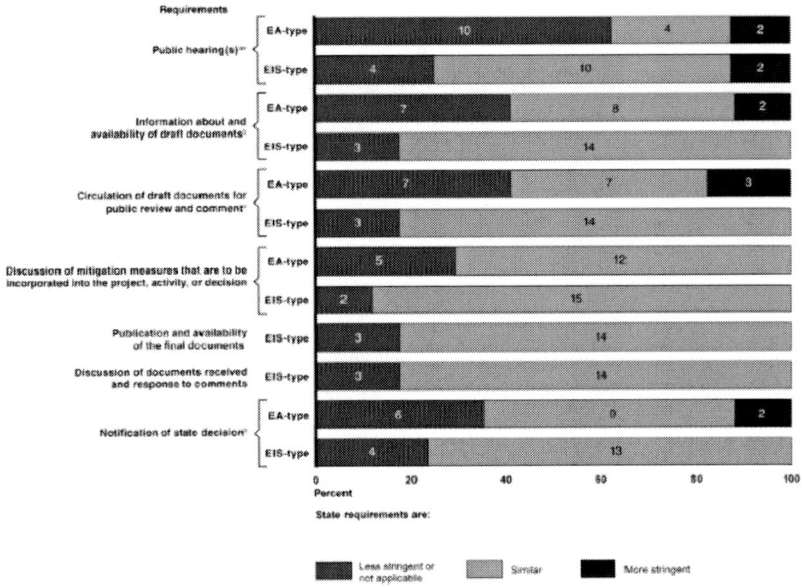

Source: GAO analysis of survey results. | GAO-15-17.

[a] For the question on requirements for public hearings there was one state that did not respond.

[b] Five states account for the eleven instances in which states reported more stringent requirements. The five instances in which three or fewer states reported less stringent requirements are divided among three states—the District of Columbia, Georgia, and New Jersey.

Figure 13. Comparison of State Requirements with Federal Highway Administration (FHWA) National Environmental Policy Act (NEPA) Requirements for Public Involvement for Environmental Assessments (EA) and Environmental Impact Statements (EIS), as Indicated by State Departments of Transportation (state DOT) Survey Responses.

FHWA's NEPA requirements allow for robust public involvement in the NEPA process, requiring reasonable notice of and an opportunity to participate in public hearings, where appropriate, and an adequate and meaningful opportunity to submit comments. State responses to our survey indicated that public involvement requirements vary, ranging from states that have no requirement to allow public involvement, to others that may have more stringent requirements than FHWA's NEPA rules.[64] Moreover, we found as part of our legal analysis that requirements allowing public participation for state EIS-type reviews are more likely to parallel FHWA's NEPA requirements than do such requirements for state EA-type reviews. Conversely, public involvement requirements for EA- type reviews in Wisconsin may exceed FHWA's NEPA requirements in some circumstances as the state's SEPA allows for additional hearings by request, while New York officials told us that they are not required to conduct a public hearing for either EA-type or EIS-type reviews. In several other states, hearings for EIS-type reviews are not automatic.[65]

Individual state requirements for notification and circulation of draft documents for comment also vary, as do requirements for public hearings. (See figure 13.) While only 3 (of 17) states surveyed reported less stringent requirements for disseminating draft EIS-type documents than required by FHWA NEPA, 7 (of 17) states surveyed reported less stringent requirements for draft EA-type documents. By contrast, Massachusetts state officials reported having more stringent requirements for providing and circulating draft EA-type documents because the state posts environmental review documents on a public website and requires a written response to all comments by the lead agency. Some states have less stringent requirements governing public involvement, particularly regarding public hearings for EA-type reviews in which 10 (of 17) states surveyed reported less stringent requirements. In our legal analysis, we found that while Indiana requires EIS-type reviews to be made publicly available, it does not require the transportation agency to seek or respond to public comments on draft versions of these documents, and Virginia's law contains no specific public notice and comment procedures for environmental review documents.

States with less stringent formal public participation requirements may in practice align with FHWA. Officials in three states told us that they match the higher standard of FHWA's NEPA public involvement requirements for state-only reviews to meet public expectations, even if less was required by state law. According to state officials in North Carolina, for example, this could mean holding a public hearing and addressing comments for a state-only EA when neither step is required by state law. According to New York state DOT officials, they conduct public hearings for EA- and EIS-type reviews, even though state law is less rigorous than NEPA and does not require hearings. Additionally, some state officials we interviewed reported practices that served to encourage public involvement, such as Washington's on-line registry of ongoing and completed reviews, which allows citizens or groups to search for projects by location.[66]

FEW INSTANCES OF FEDERAL–STATE DUPLICATION WERE REPORTED; FHWA AND STATES HAVE IMPROVED COORDINATION

Few States Reported Duplication Resulting from Federal and State Environmental Review Processes; More Than Half Reported No Duplication for Any Reason

Officials in 4 (of 18) states expressly identified instances of federal–state duplication in environmental review processes. For purposes of this report, we focused on duplication that might occur between state and federal processes where there is duplication of effort.[67] In those instances where state officials identified duplication, it resulted either from supplemental state requirements or from the misalignment of federal and state environmental review documents, according to state officials. Officials in Washington reported duplication from both of these causes.

- Two of the four states reporting potential federal–state duplication in our survey, Maryland and Washington, include a SEPA EA-checklist in addition to their federal review documents. State officials in both states explained that completing the checklist may be duplicative because it includes information that is similar to the information in the FHWA review, but noted that doing so is not burdensome in terms of time or resources. More specifically, in Washington, state officials said that the checklist serves a “due diligence” role for the state’s SEPA, while Maryland officials said the checklist is used to scope EIS-type reviews during the beginning of the review process.
- Three states we surveyed, including Washington, reported that the lack of alignment between required state and federal document types for the same project could result in additional effort. For example, both Massachusetts and Minnesota sometimes use a different state review type for some projects than would be required for the FHWA NEPA process, depending on characteristics of the project. Consequently, in Minnesota, some projects are reviewed with a state

EA-type process, while being categorically excluded by FHWA. In these situations, parallel efforts may be required to satisfy the different requirements— which may be more stringent for one of the required reviews— precluding the use of a blended process or combined document to coordinate similar processes. Washington has recently completed a rulemaking to align federal and state CEs to avoid this type of duplication.

Ten (of 18) states responding to our survey reported that there was no duplication in either the procedural steps or substantive tasks required by state or federal environmental review requirements.[68] In seven of these states, adopting a NEPA review fulfills SEPA requirements, and state officials pointed to this adoption as the reason for reporting no duplication in their survey responses.[69] Connecticut and Hawaii allow for the adoption of the federal NEPA review as long as the review meets certain state SEPA requirements, according to state officials. Montana and New York reported that their processes are not duplicative because they have integrated or blended processes to meet both sets of requirements concurrently. According to state and federal officials we interviewed, several other reasons contribute to minimal duplication between state and federal processes. The development of required state and federal environmental review documents is typically carried out by the same state officials (or other project sponsors), who can use analyses for different purposes without replicating effort when federal and state requirements are similar. Finally, state and federal review processes frequently require—or encourage—coordination, as mentioned above, and officials pointed to this coordination as a reason for the lack of duplication. For example, officials in North Carolina described meetings held every 2 months involving state and federal officials from the transportation and resource agencies that are typically involved in environmental review. At these meetings, officials are able to coordinate state and federal efforts, among other things.

We previously found that FHWA does not collect information on the cost of environmental reviews, and during the course of this review, FHWA officials at the headquarters and division levels confirmed that such data are not collected.[70] Likewise, state officials we surveyed reported no information on the cost to states of any federal–state duplication, and, as mentioned above, officials in those states identifying instances of federal– state duplication from the use of SEPA checklists described the additional effort as being negligible, although they were unable to quantify any additional costs.[71]

States Reported Other Examples of Duplication or Overlap

During the course of our review, we identified other examples of potential duplication or overlap, but these did not result from the interaction of federal and state environmental review requirements. In addition to the four states reporting federal–state duplication, four additional states reported other examples of potential duplication or overlap.[72] For example, state DOT officials in Maryland noted that there may be potential duplication resulting from the need to keep federal reviews up to date for projects taking a number of years, either because initial reviews have expired or because environmental conditions—or requirements— have changed.[73] In our survey, three states reported potential duplication—or overlap—when federal permitting agencies (e.g., the Corps) carried out additional analyses because they did not accept the FHWA-approved NEPA review. CEQ officials noted, however, that while the goal of the NEPA process may be that there is one NEPA review and approval process, often the level of detail required by one agency (e.g., FHWA) to review a proposed decision under NEPA may be different than what is required by another agency to issue a permit (e.g., the Corps or the Coast Guard).

Officials with Caltrans reported potential duplication in our survey, but did not provide examples of the cause or type of duplication. When we interviewed Caltrans officials, they explained that while additional analyses may be required for the state's SEPA, there is no duplication of effort caused by the interaction of the state and federal requirements given the agency's blended review process. Separately, officials with the California State Association of Counties contacted us regarding potential duplication in the development of state and federal reviews at the local level. Under the state's SEPA, local governments prepare both state and FHWA NEPA environmental reviews for local projects, but they can approve only the SEPA reviews. FHWA or Caltrans has approval authority for FHWA NEPA reviews.[74] County officials stated that they are not able to do a blended document given the different reviewers, a situation that results in potential duplication, adding time and additional cost to projects.[75] We spoke with officials with the National Association of Counties, and they stated that this concern has not been raised by county officials outside of California.

States Reported Benefits from Efforts to Improve Coordination and Efficiency, Including Reduced Uncertainty in Projects' Time Frames

All 18 states we surveyed reported having agreements with FHWA or other federal agencies to improve coordination or make environmental review processes more efficient. Nearly all of these states (16 of 18) provided examples of programmatic agreements, such as those allowing state officials to review and approve CE determinations for FHWA or other efforts to improve interagency coordination for conducting reviews and obtaining permits.[76]

Many programmatic agreements and other improvement efforts were developed as part of FHWA's Every Day Counts initiative,[77] which is an effort to identify and deploy innovation aimed at shortening project delivery, enhancing the safety of roadways, and protecting the environment.[78] Examples of FHWA or state-led improvement efforts identified by state officials we surveyed and interviewed include the following:

- Two states in our survey have developed an interagency process to plan and review highway projects under the auspices of the Section 404 merger process to reduce inefficiencies in assessment and permitting under the Clean Water Act. For example, officials in North Carolina told us that the state has been working to make environmental reviews more efficient through its Section 404 merger process since the 1990s, through which the state DOT coordinates with FHWA, the Corps, the U.S. EPA, and several other state and federal agencies, including the state's Department of Environment and Natural Resources.
- Three states in our survey reported having liaison positions with other state or federal agencies to alleviate resource challenges and to improve interagency coordination. For example, Texas funds a liaison position at U.S. Fish and Wildlife Service. Massachusetts also funds such positions with the Corps—in collaboration with FHWA—and the Massachusetts Departments of Environmental Protection and Fisheries and Wildlife. According to Massachusetts DOT officials, having a liaison with the Corps has facilitated coordination, reduced the time needed to review permit applications, and allowed for the use

of more proactive protection for some endangered species, including turtles.[79]

Officials in two other states, California and Wisconsin, also pointed to improved coordination by funding positions in coordinating agencies.

- Eleven (of 18) states in our survey reported having programmatic agreements under Section 106 (assessing impacts on historic properties) with FHWA and other federal agencies. According to FHWA, Section 106 programmatic agreements are one way to expedite the environmental review process, while protecting and enhancing the environment. These agreements authorize state DOTs to conduct all or some Section 106 reviews on behalf of FHWA, when such reviews are required.
- Individual states have efforts to improve processes, as well. For example, officials with the California Office of Planning and Research worked with CEQ to develop a handbook on the interaction of state environmental review requirements and NEPA to smooth and better coordinate the dual reviews that are often required, which was released in 2014.

  According to a state official, many of the challenges in coordinating the two processes stemmed from a lack of understanding of the other requirements, and the state worked with CEQ to develop a guide to explain key differences and to define terminology.

Eleven (of 18) states responding to our survey reported benefits from efforts to improve coordination or to make state and federal processes more efficient, most notably decreased time frames. Other benefits included increased public involvement, increased agency engagement, and decreased costs.

State officials were unable to quantify these benefits. State officials also pointed to increased certainty in project timelines, costs, and processes, as well as improved coordination with other agencies and tribes.

AGENCY COMMENTS

We provided a draft of this report to the U.S. Department of Transportation (DOT) and the Council on Environmental Quality (CEQ) for review and comment.

DOT and CEQ provided technical corrections about federal and state environmental review requirements, which we incorporated as appropriate.

David J. Wise,
Director, Physical Infrastructure Issues

Susan Sawtelle,
Managing Associate General Counsel

Appendix I. A Comparison of Federal NEPA Requirements and State Environmental Policy Act Requirements as Applied to Federal-Aid Highway Projects

As discussed in the body of this report, we identified state statutes, regulations, and orders in the 18 states where review of highway projects is required under a state environmental policy act (SEPA), and we compared those requirements to the federal requirements for federal-aid highway projects under the National Environmental Policy Act (NEPA) and implementing regulations issued by the Council on Environmental Quality (CEQ) and the Federal Highway Administration (FHWA). We focused our comparison on 12 key elements characterizing NEPA programs, which we developed in consultation with CEQ and FHWA. This appendix summarizes the results of our review of state legal requirements, not state practices, as discussed in the body of this report, and the examples are illustrative and are not intended to describe all aspects of each state's SEPA program.

Element 1: Policy and Purpose of the Environmental Review Requirements

The key purposes of NEPA include:

- declaring a national policy which will encourage productive and enjoyable harmony between man and his environment; and
- promoting efforts which will prevent or eliminate damage to the environment and biosphere and stimulate the health and welfare of

man and enriching understanding of the ecological systems and natural resources important to the nation.

42 U.S.C. § 4321.

A majority of SEPAs establish objectives that are at least somewhat similar to these NEPA goals, although a few states have more limited policy goals.[80] Many states' SEPAs establish a state policy that encourages harmony between humans and their environment or that enriches understanding of the natural environment.[81] In addition, SEPAs generally establish a state policy requiring that environmental concerns be evaluated in connection with state-funded or otherwise state- supported projects,[82] although several do not.[83]

Several SEPAs establish a policy of avoiding or mitigating environmental damage. Others explicitly refer to the desirability of informing and involving the public in environmental decision making, and some SEPA purpose-and-policy statements specifically refer to management of natural resources, waste disposal or maintenance of the public health. A few SEPAs address local functions such as land use management and zoning. Finally, a few SEPAs have other declared objectives such as strengthening the state economy (Georgia, *see* GA. CODE ANN. § 12-16-2(1)) or supporting the right to use and enjoy private property (Montana, *see* MONT. CODE ANN. § 75-1-102(2)).

Element 2: Types of Projects Covered

NEPA generally requires review of "every recommendation or report on proposals for legislation and other major Federal actions significantly affecting the quality of the human environment." 42 U.S.C. § 4332(2)(c). Our review looked at how NEPA applies to projects authorized under Title 23 of the U.S. Code, consisting primarily of federal-aid highway projects where FHWA provides funds to state or local governments and serves as the federal lead agency for the NEPA review. FHWA's review also includes supplemental environmental requirements contained in Title 23.

When a project is funded solely with state or local funds, it rarely requires an FHWA NEPA review. Federal permitting requirements also can trigger a NEPA review, however, even when only state or local funds are used such as when U.S. Army Corps of Engineers (Corps) or U.S. Coast Guard (Coast Guard) approval is required due to the presence of a wetland or navigable waterway. In those cases, the Corps or Coast Guard could serve as the federal lead agency.

SEPAs vary with respect to which state or local "actions" trigger review. The California environmental assessment statute, for example, generally applies to any activity of any public agency that may have a substantial environmental impact, including highway projects. "Public agencies" include state agencies, boards and commissions, as well as local agencies including counties, cities, regional agencies, public districts, redevelopment agencies and other political subdivisions. CAL. PUB. RES. CODE § 21062. "Projects" include activities undertaken directly, financed in whole or in part, or requiring approval by a government agency and are generally any activity subject to the state statute. CAL. PUB. RES CODE. § 21065. Certain projects in California have been excluded from review, however. *See, e.g.*, CAL. PUB. RES. CODE. §§ 21080.01-21080.14.

By comparison, Massachusetts' SEPA covers all projects undertaken or financially assisted by a government agency, including any authority of any political subdivision. MASS. GEN. LAWS, ch. 30, §§ 61, 62. Minnesota requires an environmental impact study where there is potential for significant effects from major government actions, defined to include local and municipal agencies. MINN. STAT., § 116D.04, Subd. 1a(e). Some states' SEPAs do not apply to locally managed and funded projects, for example Georgia (GA. CODE ANN. § 12-16-3(5)) and Maryland (Md. CODE ANN., NAT. RES. § 1-301(e)). Maryland's SEPA also focuses primarily on providing environmental assessments to the state legislature. Md. CODE ANN., NAT. RES. § 1-301(d). One state's SEPA—North Carolina's—does not apply to local units of government unless they elect to be covered. *See* N.C. GEN STAT. § 113A-8(a).

Finally, SEPAs vary with respect to which project characteristics trigger an environmental review, as well as what type of review is required, and can include thresholds related to project costs and physical length, project use, and geographic area, among other things. These thresholds differ from the federal triggers for the type of review, which generally focus on the potential for significant environmental impacts rather than the scale or size of the project.[84] For example, in Virginia, reviews are required for state projects costing $500,000 or more. VA. CODE ANN. § 10.1-1188(A). Other dollar-value thresholds are built into the SEPA requirements of the District of Columbia[85], Georgia,[86] and New Jersey.[87] Some states, such as Massachusetts, require environmental impact reviews for the construction of new roadways 2 or more miles in length. 301 MASS. CODE REGS., tit.301, § 11.03(6)(a). Likewise, Minnesota requires an Environmental Impact Statement (EIS) for new road projects four or more lanes in width and 2 or more miles in length. Minn. R. 4410.4400, Subp. 16. Minnesota also requires certain types of reviews for new roads over 1 mile in length that will function as collector roadways. Minn. R. 4410.4300, Subp. 22. New York's SEPA does not apply to projects within the jurisdiction of the Adirondack Park Agency, N.Y. COMP. CODES R. & REGS., tit. 17, § 15.2(l)(3), and no EIS is required under the District of Columbia's SEPA if the project is located in what is known as the Central Employment Area, an area including but not limited to federal government facilities. D.C. CODE § 8-109.06(a)(7). Finally, some states, such as Indiana, do not extend coverage to state licensed or permitted projects, *see, e.g.,* IND. CODE § 13-12-4-8, while Texas' SEPA applies only to certain transportation projects.[88]

Element 3: Level of Detail ("Depth") of Environmental Impact Evaluation

Under NEPA, an EIS must be prepared for a project that has the potential for a "significant" effect on the environment.[89] "Categorical exclusions" (CEs) apply to projects fitting within a category of activities previously determined not to have the potential for significant environmental impacts. When project effects are uncertain, an "environmental assessment" (EA) is prepared to determine whether the project may have a potentially significant impact on the human environment. An EA briefly provides evidence and analysis sufficient to determine whether to prepare an EIS or a "finding of no significant impact" (FONSI). A FONSI presents the reasons why the agency has concluded that no significant environmental impacts will occur if the project is implemented.

A majority of SEPA states have adopted processes that provide for analyses that are generally comparable to the federal approach. Connecticut, for example, requires an EIS-type report if there are effects that "could have a major impact on the state's land, water, air, historic structure and landmarks, existing housing, or other environmental resources, or could serve short term to the disadvantage of long term environmental goals." CONN. GEN. STAT. § 22A-1C. Wisconsin's law, like NEPA, requires an EIS for "major actions significantly affecting the quality of the human environment," *see* WIS. STAT § 1.11(2)(c), and its implementing regulations identify four types of analysis and specify numerous types of transportation projects for which each type of analysis must be performed.[90] Texas's environmental review process for highway projects is by definition similar to NEPA because Texas regulations defer to FHWA's procedures whenever there would otherwise be any inconsistency between Texas' and FHWA's processes.[91] *See* 43 TEXAS ADMIN. CODE § 2.84(f). The Massachusetts SEPA treats any damage to the environment as significant, excluding only that which is found to be "insignificant." MASS. GEN. LAWS, ch. 30, § 61.

Finally, several SEPA states leave the decision whether to prepare an EIS and the extent of any EIS largely to the discretion of state project management officials. Virginia determines environmental effects using what it calls a "Preliminary Environmental Inventory," a computer- generated summary of environmental features derived from state databases and submitted to resource agencies. The agency project manager receives this information and may or may not prepare an EIS.[92]

Element 4: Level of Significance ("Breadth") Addressed in Environmental Impact Evaluation

NEPA and its implementing regulations require consideration of the significance of a project's direct, indirect, and cumulative effects. Direct effects are those "caused by the action and occur at the same time and place." 40 C.F.R. § 1508.8(a). Indirect effects are the secondary consequences on local or regional social, economic or natural conditions or resources which could result from additional activities (such as associated investments and changed patterns of social and economic activities) induced or stimulated by the proposed action, both in the short- term and in the long-term. 40 C.F.R. § 1508.8(b). Cumulative impacts are the impacts on the human and physical environment which result from the incremental impact of the proposed action when added to other past, present or reasonably foreseeable future action. 40 C.F.R. § 1508.7. Positive as well as negative impacts, and long-term as well as short-term impacts, must be considered. For federal-aid highway projects, Title 23 and FHWA also require consideration of potential project impacts on certain types of pubic parklands and historic sites, *see* 23 U.S.C. § 138, 49 U.S.C. § 303 (so-called "section 4(f) requirements").[93]

Like NEPA, several SEPA states and jurisdictions, such as Massachusetts (MASS. GEN. LAWS, ch. 30, §§ 61, 62A), Minnesota (MINN. R. 4410.1700, Subp 7), New York (N.Y. COMP. CODES R. & REGS., tit. 6, § 617.7) and Puerto Rico (P.R. REGS. JCL REG. 7948, Rule 109 DD), require broad consideration of a project's impacts. Requirements or practices in a number of states differ from NEPA requirements, however. For example, about half of the SEPA states limit consideration of indirect or cumulative impacts, the presence of which may increase the need for an EIS (their presence alone does not require an EIS under NEPA). In Connecticut, for example, cumulative effects do not need to be considered if they are not caused by the lead agency or the proposed project. *See* CONN. AGENCIES REGS. § 22a-1a-3(b) (cumulative impacts result from the incremental impact of the action when added to other past, present or reasonably foreseeable future actions "to be undertaken by the sponsoring agency"). *See also* Ga. CODE ANN. § 12-16-8(3); 326 IND. ADMIN. CODE § 16-2.1-6(5); MONT. ADMIN. R.18.2.238(1). In addition, several states only consider environmental justice or economic impacts if they have a direct impact on physical conditions within the area affected by the project. For example, while the California statute requires consideration of "growth-inducing impacts," the law also states that "evidence of social or economic impacts" can only be shown by evidence establishing "a physical impact on the environment." Cal. Pub. Res. § 21082.2(c). Nearly half of SEPA states (the District of Columbia,[94] Georgia,[95] Hawaii,[96] Indiana,[97] Maryland,[98] New Jersey[99], Virginia[100], and Wisconsin[101]) have statutes, executive orders and regulations that include little or only general discussion of indirect or cumulative impacts.

Element 5: Consideration of Alternatives

NEPA requires an EIS to contain a detailed statement regarding "alternatives to the proposed action." 42 U.S.C. § 4332(2)(c)(iii). The agency must rigorously explore and objectively evaluate all "reasonable" alternatives to the proposed action, including a "no action" alternative, in response to a specified underlying purpose and need. 40 C.F.R. §§ 1502.13, 1502.14; 23 C.F.R. §§ 771.123(c), 771.125. *See generally Biodiversity Conservation Alliance v. Jiron,* 762 F. 3d 1036 (10th Cir. 2014). NEPA does not specifically require agencies to choose the most environmentally protective alternative, or indeed any specific alternative. FHWA policy provides that federal-aid highway decisions should be made "in the best overall public interest based upon a balanced consideration of the need for safe and efficient transportation; of the social, economic, and environmental impacts of the proposed transportation improvement; and of national, State, and local environmental protection goals." 23 C.F.R. § 771.105(b).

Most SEPAs also require the relevant agency (typically the state transportation agency) to analyze the environmental impacts of alternatives to the proposed project, in addition to impacts of the proposed project itself, and most require inclusion of a no-action alternative. Many SEPAs do not specify in detail how alternatives should be evaluated, although some states specify the types and characteristics of the alternatives that must be considered or not considered. For example, Minnesota regulations require the agency to consider alternative sites, technologies, and modified designs or layouts in preparing EISS. MINN. R. 4410.2300G. Many states, like NEPA, also require consideration only of reasonable or feasible alternatives.[102]

A few states favor selection of a particular alternative or prohibit selection of certain options. The California legislature, for example, has declared that "it is the policy of the state that public agencies should not approve projects as proposed if there are feasible alternatives . . . available which would substantially lessen the significant environmental effects of such projects." CAL. PUB. RES. § 21002. The District of Columbia prohibits selection of an alternative that would substantially endanger public health, safety, or welfare, unless those effects can be avoided or mitigated. D. C. CODE § 8-109.04. Minnesota requires selection of any "feasible and prudent alternative consistent with the reasonable requirements of the public health, safety, and welfare." 11D MINN. STAT. ANN. § 116D.04, subdiv. 6. And in Wisconsin, the agency must select the alternative that is in the best overall public interest, determined by a balanced consideration of several factors including the findings of the EIS and the need for a safe and efficient transportation system. WIS. ADMIN. CODE TRANS. § 400.06(3).

Element 6: Consideration of Mitigation

NEPA and its regulations require agencies to consider mitigation of adverse environmental impacts in some circumstances, but do not specifically require agencies to carry out mitigation. Mitigation is defined to include: (a) avoiding the impact altogether by not taking a certain action or parts of an action; (b) minimizing impacts by limiting the degree or magnitude of the action and its implementation; (c) rectifying the impact by repairing, rehabilitating, or restoring the affected environment; (d) reducing or eliminating the impact over time by preservation and maintenance operations during the life of the action; and (e) compensating for the impact by replacing or providing substitute resources or environments. 42 U.S.C. § 4332(2)(C)(ii); 40 C.F.R. § 1508.20. FHWA requirements, by contrast, require reasonable mitigation measures to be taken, which are eligible for federal funding. 23 C.F.R. § 771.105(d).

Likewise, many SEPAs require that a project's environmental review documents identify mitigation measures that could lessen the environmental effects of a project. For example, in Wisconsin, a project's Record of Decision

(ROD) must indicate that all practicable means to avoid or mitigate environmental harm have been adopted or, if not adopted, include a statement explaining why. WIS. ADMIN. CODE TRANS. § 400.04(23). The ROD also must identify "mitigation measures selected" or the "reason for rejection of suggested reasonable mitigation measures."

Other state SEPA mitigation requirements vary. For example:

- The California legislature has declared that "it is it is the policy of the state that public agencies should not approve projects as proposed if there are . . . feasible mitigation measures available which would substantially lessen the significant environmental effects of such projects." CAL. PUB. RES. CODE § 21002. *See also* CAL. PUB. RES. CODE § 21100(b)(5); CAL. CODE REGS., tit. 14, § 15041 (authority to mitigate); *cf.* CAL. CODE REGS., tit. 14, § 15004(b) (prohibiting actions that would adversely affect or limit the viability of mitigation measures). Massachusetts and New York have similar requirements.[103]
- The District of Columbia does not authorize approval of a project that would have a significant environmental effect unless mitigation measures are available that would reasonably eliminate the adverse effects. In particular, if the EIS identifies an adverse effect that would substantially endanger the public, the District government must disapprove the action unless the applicant proposes mitigating measures to avoid the danger. D.C. CODE ANN. § 8-109.04.
- Some states link or combine identification of mitigation measures with identification of alternatives. Hawaii's law, for example, requires mitigation to be considered as part of its alternatives analysis, but only if mitigation actions are proposed. HAW. REV. STAT. § 343-2 (EIS must include "measures proposed to minimize adverse effects"). Moreover, Hawaii officials told us that environmental mitigation actions, once considered, are not binding. Montana's statute, by comparison, defines its alternative analysis to include mitigation, *see* MONT. CODE ANN. §§ 75-1-102(2), 75-1-220(1), and EAs can be used where the action is one that might normally require an EIS if the effects which might otherwise be deemed significant appear to be capable of mitigation by making design changes, imposing

enforceable government controls or stipulations, or both. MONT. ADMIN. R. 18.2.237(4).

Element 7: Requirement for Collaboration to ENHANCE EFFICIENCY and Avoid Duplication

To achieve efficiencies and to minimize duplication, CEQ's and FHWA's regulations require all federal agencies to collaborate with each other, and with state and local governments, to the fullest extent possible.[104]

Collaboration begins with consultation with other relevant federal and state agencies, Indian tribes, and the public; includes early identification of stakeholders, project scoping, and project planning; and extends through development of draft and final environmental impact documentation.[105]

The regulations reflect the federal government's policy to encourage collaboration of all interested parties from the outset on projects that may require environmental impact analyses, including involvement of state agencies and other federal agencies. 23 U.S.C. § 139 codifies and expands the CEQ regulatory practices as statutory mandates for federal- aid highway projects, designating the Department of Transportation (DOT) as the federal lead agency and requiring the Secretary to administer the NEPA process, including optional establishment of a schedule for completion of the environmental review process. The Secretary is also encouraged by statute to facilitate use of programmatic approaches through which states may be authorized to resolve issues that would otherwise require federal action.[106]

About half of SEPA states have policies that specifically promote or require collaboration. For example:

- California generally requires collaboration among lead, responsible and trustee agencies assisted by the Governor's Office of Planning and Research. CAL. PUB. RES. CODE §§ 21080.1, 21080.3, 21080.4. In this regard, the California legislature has recognized the importance of processes such as tiering to avoid duplicative analysis of environmental effects. CAL. PUB. RES. CODE § 21093.
- Minnesota requires responsible governmental units to collaborate to the extent practicable to avoid duplication of effort between state and federal environmental reviews and between environmental reviews and environmental permitting. MINN. STAT. § 116D.04, Subp. 2a(g).

- Other states with formal cooperation policies include Connecticut (CONN. GEN. STAT. § 22a-1; *see also* CONN. GEN. STAT. § 22a-1a); the District of Columbia (D.C. CODE § 8-109.07); New York (N.Y. COMP. CODES R. & REGS., tit. 6, §§ 617.3(d), 617.6; N.Y. ENVTL. CONSERV. LAW § 8-0111.1); and Montana (through its rules, which include a section on cooperation with federal agencies, MONT. ADMIN. R. 18.2.250).

The law in some states is unclear regarding how broadly state agencies are required to cooperate with federal agencies, including federal resource agencies. For example, although the Hawaii SEPA lists cooperation and coordination as important government objectives, *see* HAW. REV. STAT. § 343-1, the regulations refer specifically to the importance of cooperation and coordination between the state accepting authority or approving agency and other state authorities or agencies only in determining the applicability of requirements for supplemental environmental statements (*see* HAW. ADMIN. CODE § 11-200-27) and in avoiding duplication with NEPA requirements. The North Carolina regulations authorize but do not require state agencies to seek information from federal as well as local and special units of government. 1 N.C. ADMIN. CODE § 25.0210. Puerto Rico requires consultation with federal and state agencies prior to submitting the environmental document but does make the recommendations of federal agencies within their areas of jurisdiction binding. P.R. Reg. 7948, Rule 118 E.

Finally, New York requires state cooperation with federal agencies. N.Y. ENVTL. CONSERV. LAW. § 8-0111.1. Washington establishes as policy that the Department of Ecology is to "utilize, to the fullest extent possible, the services, facilities, and information (including statistical information) of public and private agencies, organizations, and individuals, in order to avoid duplication of effort and expense, overlap, or conflict with similar activities authorized by law and performed by established agencies." WASH. REV. CODE § 43.21C.110(2)(b). Wisconsin incorporates CEQ's (but not FHWA's) processes, (*see* WIS. ADMIN. CODE TRANS. § 400.06(b), App. citing 40 C.F.R. § 1500.5), while North Carolina authorizes but does not require that agencies seek information from federal as well as local and special units of government. (*see* 1 N.C. ADMIN. CODE § 25.0210(b)).

Element 8: Requirement for Coordination to Ensure Involvement of All Relevant Agencies

Title 23 and CEQ regulations require the lead federal agency to coordinate the timing and scope of its reviews with cooperating agencies. 23 U.S.C. § 139(g); 40 C.F.R. § 1501.7(a)(6). Generally, for Title 23 funded projects, the lead federal agency is a modal administration within the Department of Transportation. *See* 23 U.S.C. § 139(c). *See also* 40 C.F.R. §§ 1501.1(b) (early and cooperative interagency consultation); 1501.2(d)(2) (requiring federal agency consultation with state, local, and tribal authorities and private persons and organizations); 1501.5 (lead agencies); 1501.6 (cooperating agencies).

Some states' SEPAs also provide for robust coordination. For example:

- California generally requires not only collaboration (as discussed in Element 7 above) but also requires coordination among lead, responsible and trustee agencies assisted by the Office of Planning and Research. *See* CAL. CODE REGS., tit. 14, § 15082(c). And when a proposed project is of sufficient statewide, regional, or area-wide environmental significance, California uses a clearinghouse process to facilitate and coordinate review of draft Environmental Impact Reports and other environmental documentation. CAL PUB. RES. CODE, §§ 21083(d) (review of draft EIRs, negative declarations, or mitigated negative declarations); CAL. PUB. RES. §§ 21082.1(c) (4), 21091. *See also* CAL. CODE REGS., tit. 14, §§ 15004(b) (timing), 15006 (reducing delay and paperwork), 15083 (consulting), 15063(c)(2) (mitigating effects to facilitate documented CE or negative declaration).
- Hawaii requires that "[t]he [Office of Environmental Quality Control within the state Department of Health] and agencies shall cooperate with federal agencies to the fullest extent possible to reduce duplication between federal and state requirements." Haw. Rev. Stat. § 343-5(h).
- In Massachusetts, the SEPA review process "is intended to involve any interested Agency or Person as well as the Proponent and each Participating Agency." Code of Mass Regs., tit. 301, § 11.01(b).
- In Minnesota, to the extent practicable, responsible governmental units must avoid duplication and ensure coordination between state and federal environmental review and between environmental review and environmental permitting. Minn. Stat. § 116D.04, Subd. 2a(g).

In other states, the laws require little or no formal attention to coordination in applying their SEPA requirements. For example:

- The Maryland MDOT regulations only require that the lead agency describe "the coordination and liaison relationship established in developing the proposal," with the content of the description largely up to the agency. There is no clear requirement defining the responsibilities that the lead agency assumes. 11 CODE OF MD. REGS. § 01. 08.03(B)(9). *See also* 11 CODE OF MD. REGS. § 01.08.04(A)(3) ("The timing and type of community and public agency involvement in this analysis will be determined on a case-by-case basis . . ..").
- Virginia provides specifically for coordination between VDOT and state resource agencies only. VA. CODE ANN. § 10.1-1191.
- Coordination is not mentioned in the New Jersey executive order, N.J. GOV. KEAN, EXEC. ORDER NO. 215, or in guidance.

Element 9: Requirement for Single-Process Review

Reflecting the requirements for federal coordination and collaboration discussed in Elements 7 and 8 above, federal environmental reviews of Title 23-funded highway projects also must include "completion of any environmental permit, approval, review, or study required for a project under any Federal law other than NEPA." 23 U.S.C. § 139(a)(3)(B). This requirement reflects a policy that the NEPA process and the permitting processes should not be treated as separate and distinct processes but as one.

Most SEPA states do not require or conduct single-process reviews as such. As discussed in Element 8 above, the Minnesota statute requires that the responsible governmental unit shall, to the extent practicable, avoid duplication and ensure coordination between state and federal environmental review and between environmental review and environmental permitting. MINN. STAT. § 116D.04, Subp. 2a(g). North Carolina and Washington require a list of all licenses that the project is known to require, *see* 1 N.C. ADMIN. CODE § 25.0603(2), WASH. ADMIN. CODE § 197-11-440(2)(d), and Washington has made significant efforts to integrate its Growth Management Act and Model Toxics Control Act processes with its SEPA processes.[107] In Puerto Rico the responsibility for issuing construction permits is centralized in a Permit Management Office. *See* P.R. LAWS ANN., tit.12, § 8001a(c). This office assesses compliance with Puerto Rico's Environmental Public Policy

Act. *See id.* Finally, as noted in Element 3 above, Texas regulations defer to FHWA's procedures whenever there would otherwise be any inconsistency between Texas' and FHWA's processes.

Element 10: State Adoption of Federal NEPA Reviews

To avoid duplication that might otherwise result, a number of SEPAs authorize or encourage preparation of documentation that meets both federal and state NEPA and SEPA requirements, or use of information, documentation, or analyses developed for the NEPA review. The SEPA procedures vary from state to state, ranging from provisions allowing use of some or all of the paperwork prepared to meet federal requirements to full adoption of the federal process and results so that no separate state funds are required. These issues arise in the context of three basic scenarios:

- About half of SEPA states are authorized to forgo the SEPA process and determination entirely if the proposed project is covered by a completed NEPA review. For example, in Georgia, an agency is deemed to have complied with the requirements of the SEPA if the proposed government action requires and has received federal approval of an environmental document prepared in accordance with NEPA. *See* GA. CODE ANN. § 12-16-7. In Indiana, if any state agency is required by NEPA to file a federal EIS, it is not also required to file an EIS with the state government. *See* IND. CODE § 13-12-4-10.
- Some SEPA states are authorized to use NEPA documentation to meet their SEPA requirements but state officials must make some kind of independent decision under state law. For example, in Minnesota, if a federal EIS will be or has been prepared for a project, the state may use the draft or final federal EIS as the draft state EIS if the federal EIS addresses the scoped issues and satisfies the state content standards for an EIS. *See* Minn. R. 4410.3900, Subp. 3; Minn. R. 4410.2300. In Montana, implementation of NEPA and the Montana SEPA are separate and distinct federal and state functions, but state agencies are required to coordinate with other state and federal agencies in the preparation of a single environmental review that is legally sufficient for both NEPA and MEPA. MONT. ADMIN. R. 18.2.250(c).

- Some SEPA states allow preparation of a single set of documentation meeting both the NEPA and additional SEPA requirements.[108] The state must only prepare separate findings—in effect, a separate Record of Decision or similar documentation. For example, in California, the SEPA and regulations mandate use of NEPA EISs and other documentation in lieu of state Environmental Impact Reports and meeting other state requirements whenever possible. *See* CAL. PUB. RES. §§ 21083.5-21083.7; CAL. CODE REGS., tit. 14, §§ 15063, 15082, 15110, 15127, 15220-15528. The state SEPA does not, however, dispense with the need to meet its state-specific requirements. In New York, the SEPA requires cooperation between state and federal agencies in creating an environmental review and exempts a project from additional state review if a federal NEPA review is conducted. *See* N.Y. ENVTL. CONSERV. LAW § 8-0111(1)-(2); N.Y. COMP. CODES R. & REGS., tit 17, § 15.10. If the proposed action is subject to NEPA, the statute is interpreted to require that NYDOT must comply with the federal requirements which then excuses further statutory obligations. *See* N.Y. COMP. CODES R. & REGS. tit. 17, § 15.6; *see also* N.Y. ENVTL. CONSERV. LAW § 8-0111(1)-(2) (single combined document prepared along with state and federal report.). In Washington, state documentation is not required if federal documentation already has been prepared for the same project, *see* WASH. REV. CODE § 43.21C.150, but the statute does not waive the requirement for a state decision concerning the adequacy of any prior NEPA review, even if that decision is based on NEPA documentation. *Id.; see also* WASH. ADMIN. CODE §§ 197-11-600, 197-11-630.

Element 11: Opportunity for Public Participation

The federal NEPA process requires the opportunity for robust public participation. At the least, the public must be notified of the proposed project and given an opportunity to comment on it. *See, e.g.,* 40 C.F.R. §§ 1501.4(b), (e) (EAs); 1502.19 (EISs); 1506.6 (NEPA implementation generally). There are also specific public participation requirements in Title 23 for federal-aid highway projects. *See, e.g.,* 23 U.S.C. §§ 128, 139; 23 C.F.R. §§ 771.111(h)(1)-(2), 771.113(a)(2). Moreover, 23 U.S.C. § 128 requires that an opportunity for a public hearing be provided to consider the impact of each federal aid-highway project on the environment.

Similarly, almost all SEPAs or their implementing regulations provide for some degree of public participation. Three SEPA states do not require agencies to consider public input at all: Indiana (only authorized comments must be considered, 327 IND. ADMIN. Code §§ 11-1-4, 11-3-3), New Jersey (N.J. GOV. KEAN, EXEC. ORDER No. 215, and guidance), and Virginia (generally, comments are invited from interested agencies, planning district commissions and localities (VA Dept. of Env. Qual., Procedure Manual, July 2013)). The opportunities for public participation in other states vary. For example:

- In New York, public notice must be provided for all determinations that a project will have no significant effect; that a project may have a significant effect; that a draft or final EIS has been completed; and any subsequent notice of a negative declaration. N.Y. COMP. CODES R. & REGS. tit. 17, § 15.10. Public notice of hearings also must be given, although this may be combined with a notice of completion of a draft EIS, and public comments must be permitted for draft and final EISs. N.Y. ENVTL. CONSERV. LAW § 8-0109 (4); N.Y. COMP. CODES R. & REGS., tit. 17, §§ 15.6(d), 15.10(d). The regulations also provide for consideration of public comments on scoping and other matters. N.Y. COMP. CODES R. & REGS., tit. 6, §§ 617.8(e), 617.7.
- Washington State provides many opportunities for public participation and comment. *See, e.g.,* WASH. REV. CODE § 43.21C.110(1); WASH. ADMIN. CODE §§ 197-11-340(2)(c), 197-11-350(5), 197-11-355,197-11-400(4), 197-11-405, 197-11-408, 197-11-455(6)-(8), 197-11-502, 197-11-560. *See also* WASH. ADMIN. CODE § 197-11-410(1)(d).
- In Wisconsin, it is WisDOT's policy that "public involvement, interagency coordination and consultation, and a systematic interdisciplinary approach to analysis of the issues shall be essential parts of the environmental process for proposed actions." WIS. ADMIN. CODE TRANS. § 400.06(4). As part of the scoping process, the Wisconsin regulations "establish a schedule for document preparation and for opportunities for public involvement." WIS. ADMIN. CODE TRANS. § 400.09(4)(c). Public comment must be allowed on EISs and EAs, *see* WIS. ADMIN. CODE TRANS. § 400.11(3)-(5), but not on FONSIs and Environmental Reports. *See* WIS. ADMIN. CODE TRANS. § 400.11(6)-(7).

- In California, the importance of public participation in the SEPA process is specifically recognized in the regulations. CAL. CODE REGS, tit. 14, § 15002(a)(1), (4). Key environmental review documents are classified as public documents. *Id.*, § 15002(f), (j) ("Under [the California law], an agency must solicit and respond to comments from the public and from other agencies concerned with the project."); *see also id.* § 15022(a)(5) (duty of California public agencies to consult with the public regarding environmental effects).

A number of SEPAs limit public participation to commenting on specific aspects of the process. For example, the Massachusetts statute and regulations provide notice-and-comment procedures at critical points in the process. *See* MASS. GEN. LAWS, ch. 30, § 62C; CODE OF MASS REGS., tit. 301, § 11.15. Several jurisdictions, such as Georgia, Connecticut, and the District of Columbia, authorize public hearings but base the decision to hold them on the number of requests received. *See* GA. CODE ANN. § 12-16-5; CONN. GEN. STAT. § 22a-1d; D.C. CODE § 8-109.03 ("If 25 registered voters in an affected single member district request a public hearing on an EIS or supplemental EIS or there is significant public interest, the Mayor, board, commission, or authority shall conduct a public hearing."). Montana provides for notice of EIS scoping to "affected federal, state, and local government agencies, Indian tribes, the applicant, if any, and interested persons or groups," but not the general public. MONT. ADMIN. R. 18.2.241(2)(a).

Element 12: Opportunity for Judicial Review of Agency Decisions

The opportunity to obtain meaningful review of agency action by a court is an important protection against arbitrary, capricious or otherwise unlawful agency decision making. At the federal level, judicial review of NEPA decisions issued by lead agencies and other involved federal agencies occurs under the federal Administrative Procedure Act, 5 U.S.C. §§ 701-706 (APA).[109] Not everyone who is dissatisfied with a NEPA decision may challenge it in court; only those who suffer specific and sufficient injury as a result of the decision, and thus have "standing," may file suit. For federal-aid highway projects, a lawsuit generally must be filed within 150 days after publication of a notice in the Federal Register announcing a final approval, permit, or license. *See* 23 U.S.C. § 139(l)(1).[110]

Similarly, while some of the 18 states with SEPAs required for highway projects provide for court review of decisions in their SEPA legislation, most do so either in their general administrative agency procedure legislation (typically based on the Model State Administrative Procedure Act) or in specialized legislation. Most states also have followed federal law by limiting who may challenge the state agency decision, how the agency's action will be reviewed, and what the scope of that review will be. In particular, as applied to state and federal-aid highway projects, SEPA laws based on the Model Act[111] establish five key prerequisites to challenge a state agency decision:

- The challenger must suffer particularized harm. For example, individuals owning property that might be acquired or adjacent to the project area, or organizations representing such persons (including environmental groups), likely could bring suit.
- The agency decision must be final, that is, the challenger generally must have exhausted any dispute resolution process available at the agency.
- The court challenge generally must be based only on the evidence presented to the lead agency and the issues already raised to the agency (the administrative record).
- The challenge generally must allege that the agency's decision was legally arbitrary and capricious, contrary to state or federal law, or not supported by the evidence before the agency.
- The remedy being requested generally must be limited to an order directing the agency to take a certain action, rather than seeking monetary damages.

An important difference between federal NEPA-decision judicial review requirements and state SEPA requirements is the difference in the time for filing a challenge. The federal period is generally 150 days, as noted above, while state laws typically provide only 30 days to appeal. *See, e.g.,* IND. CODE § 4-21.5-5-5; IND. STAT § 116D.04, Subd. 10; N.C. GEN. STAT. § 150B-45. Washington provides only 21 days. WASH. REV. CODE § 43.21C.080(2)(a).

APPENDIX II: OBJECTIVES, SCOPE, AND METHODOLOGY

The Moving Ahead for Progress in the 21st Century Act (MAP-21) contains a mandate requiring that GAO review state laws and procedures for conducting environmental reviews with regard to projects funded under title 23 of the United States Code (primarily federal-aid highway projects).[112] This report addresses: (1) the factors that determine whether federal or state environmental reviews are required for highway projects, and how the types of federal and state environmental review documents compare; (2) how state environmental review requirements and practices compare with federal requirements for assessing federal-aid highway projects; and (3) the extent of any duplication in federal and state reviews, including frequency and cost, in states with environmental review requirements for highway projects.

We identified 18 states with state environmental policy acts (SEPA) required for highway projects for inclusion in our review (see table 1). In these states, statutes or regulations require some assessment of potential environmental effects from highway projects that may mirror requirements under the National Environmental Policy Act (NEPA)[113] for federal-aid highway projects. The list of states with SEPAs derives largely from the 18 states identified by the Council on Environmental Quality (CEQ) as having SEPAs, including New Jersey, which has an executive order that requires environmental reviews. In addition, during the course of our work, we learned that Texas does not have a general state-level SEPA but does have a state statute and regulations that apply to transportation projects, and we confirmed with CEQ officials that we should include Texas in our scope. By contrast, we excluded South Dakota because its SEPA provides the option of preparing an environmental impact statement, but does not require one, and South Dakota Department of Transportation officials told us that they do not conduct environmental reviews under the state law.

For each of our objectives, we reviewed relevant publications, including our prior reports on NEPA and highway projects.[114] We obtained documents and analysis from federal agencies related to NEPA reviews for federal-aid highway projects, including CEQ, the Congressional Research Service (CRS), and the Federal Highway Administration (FHWA), including FHWA's Environmental Review Toolkit for NEPA and Transportation Decisionmaking (FHWA's NEPA Toolkit), which provides guidance on FHWA's NEPA environmental review process for state department of transportation (state DOT) officials. In addition, we interviewed officials with FHWA, CEQ, and CRS. We also interviewed two academics who authored a treatise on environmental review requirements and were cited by CEQ as having expertise on NEPA and SEPAs—Professor Daniel Mandelker at Washington University and Arianne Aughey—and representatives from two professional associations with expertise in federal or state environmental review requirements or state practices—the American Association of State Highway and Transportation Officials (AASHTO) and the National Association of Counties.

Table 1. States with State Environmental Policy Acts (SEPA) Required for Highway Projects

California	Montana
Connecticut	New Jersey
District of Columbia	New York
Georgia	North Carolina
Hawaii	Puerto Rico
Indiana	Texas
Maryland	Virginia
Massachusetts	Washington
Minnesota	Wisconsin

Source: GAO. | GAO-15-71

Note: We included entities that are defined as "states" under Title 23 and that have SEPAs required to be applied to highway projects. Thus, we did not include the Nevada/California-Tahoe Region, which has SEPA requirements, because it is not defined as a state under Title 23. We included New Jersey, which has an executive order that serves as a SEPA, and Texas, which does not have a general state SEPA but does have a state statute and regulations that apply to transportation projects. We excluded South Dakota because although its SEPA provides the option of preparing an environmental impact statement, there is no requirement to do so, and South Dakota DOT officials told us that they do not conduct environmental reviews under the state law.

To respond to the first two objectives, we conducted a legal analysis and a survey, which included all 18 states that we identified as having SEPAs required for highway projects. Our legal analysis compared key elements of SEPAs and related state regulations with key elements of NEPA and FHWA regulations for federal-aid highway projects. Our comparison of NEPA and FHWA regulations with requirements in state statutes, regulations, and executive orders focused on the key NEPA statutory and regulatory requirements and did not systematically examine court decisions or legislative history. We identified the key NEPA elements by reviewing relevant federal statutes and regulations in consultation with CEQ and FHWA. Specifically, we started with the statutory language of NEPA, which requires agencies to prepare, for major federal actions significantly affecting the quality of the human environment, a detailed statement on, among other things: (1) the environmental impact of the proposed action, (2) any adverse environmental effects which cannot be avoided should the proposal be implemented, and (3) alternatives to the proposed action.[115] NEPA, CEQ regulations, and—for federal-aid highway projects—FHWA regulations then specify detailed environmental review processes, which include requirements for interagency coordination, avoiding duplication, and participation by state and local governments as well as the general public, among other things. In addition, the regulations provide for more efficient methods of environmental review and a process for determining when use of these methods is appropriate. Based on these statutory and regulatory requirements, and in consultation with CEQ and FHWA, we distilled 12 key elements of environmental review.

1. Policy and purpose of the environmental review requirements
2. Types of projects covered
3. Level of detail ("depth") of environmental impact evaluation
4. Level of significance ("breadth") addressed in environmental impact evaluation
5. Consideration of alternatives
6. Consideration of mitigation
7. Requirement for collaboration to enhance efficiency and avoid duplication
8. Requirement for agency coordination
9. Requirement for single-process review
10. State adoption of federal NEPA reviews
11. Opportunity for public participation
12. Opportunity for judicial review of agency decisions

In additional to this legal analysis, we conducted a survey of state DOTs in all the 18 states with SEPAs required for highway projects to compare in more detail state environmental review requirements with FHWA's NEPA requirements for reviews of federal-aid highway projects. Using FHWA's NEPA Toolkit as a guide, we developed survey questions to gather information comparing states' environmental review requirements and practices with FHWA's NEPA requirements.[116] These requirements align with five of FHWA's six principles of the NEPA process and reflect key elements in our legal analyses.[117]

- consideration of the social, economic, and environmental impacts of a proposed action or project;
- development and evaluation of a range of reasonable alternatives to the proposed project, based on the applicants defined purpose and need for the project;
- mitigation of project effect by means of avoidance, minimization and compensation;
- interagency coordination and consultation; and
- public involvement including opportunities to participate and comment.

For each of these principles, we developed questions to assess the extent to which state requirements were less stringent than, similar to, or more stringent than FHWA's NEPA requirements.[118] After we drafted the survey, we conducted pretests with state DOT officials in Maryland, Washington, and North Carolina to ensure that respondents interpreted our questions in the way we intended. That is, we verified that the questions were clear and unambiguous and that we used appropriate terminology in the survey, to ensure that respondents had the necessary information and ability to respond to the questions. Where necessary, we revised the screening tool to improve the survey instrument in response to feedback from the pretests and internal GAO review. We divided the final screening-tool questions into four parts: Part I: Contact Information; Part II: Documentation; Part III: Duplication and Coordination; and Part IV: Environmental Review Requirements.

We administered the survey by emailing an electronic form to state DOT officials in all 18 states with SEPAs required for highway projects.[119] Two states provided clarifications or supplementary information along with their survey responses. To improve the accuracy and completeness of the data, we used the clarifying information provided by agency officials to update responses where necessary. Because this effort was not a sample survey, it has no sampling errors. However, the practical difficulties of conducting any survey may introduce errors, commonly referred to as nonsampling errors. For example, difficulties in interpreting a particular question or sources of information that are unavailable to respondents can introduce unwanted variability into the survey results. We took steps to minimize such nonsampling errors in developing the survey tool— including using a social science survey specialist to help design and pretest the survey. We also minimized the nonsampling errors when analyzing the data, including using an independent analyst to review all computer programming related to the survey.

Finally, there were a few instances where state DOTs should have indicated one response to a question and instead provided two. In these cases we followed up with the state DOT officials to clarify their response.

To identify which states had requirements that were "generally similar" to FHWA's NEPA requirements overall, we determined which states in our survey reported having environmental review requirements that we found to be similar or somewhat similar to 42 individual requirements for FHWA NEPA reviews. We characterized state survey responses as being similar overall if 75 percent or more of the questions about individual state requirements under a NEPA principle were marked as "similar" or "more stringent" than FHWA's NEPA requirements.[120] If 50 to 74 percent of the requirements were marked as "similar" or "more stringent," we characterized state survey responses as somewhat similar. If 51 to 74 percent of the requirements were marked as "less stringent" or "not applicable," we characterized state survey responses as somewhat less stringent. If 75 percent or more of the requirements were marked as "less stringent" or "not applicable," we characterized state survey responses as less stringent. Then we determined that states where officials indicated that at least half of their responses for each requirement within the principle were similar to or more stringent than FHWA's NEPA requirements were "generally similar" to federal requirements. For example, regarding the consideration of impacts, we determined that state requirements were "generally similar" overall if the state DOT officials reported that 3 of the 5 individual requirements were at least similar (if not identical) to FHWA's NEPA requirements: (1) identification of impacts, as well as assessment of (2) cumulative effects, (3) context, (4) cultural or historical impacts, and (5) social or environmental justice. In addition to these analyses, we interviewed state DOT officials and officials with state natural resource agencies in 9 of the 18 states with SEPAs. (See table 2.) We selected SEPA states for additional interviews and site visits based on four criteria: robustness (or lack thereof), number of active EIS reviews, "uniqueness," and, in some cases, proximity to GAO offices. Our findings for these 9 states are not generalizable to the other 9 states with SEPAs but provide examples of varying state requirements and practices. To respond to the third objective, we included questions about duplication in our survey of state DOTs, as well as interviewing state officials and FHWA officials at the division offices in those states we selected for additional interviews. We have defined duplication as occurring when two or more agencies or programs are engaged in the same activities or provide the same services to the same beneficiaries, in accordance with GAO's body of work on duplication in the federal government.[121] This report focuses on duplication that might occur between state and federal processes for environmental review of highway projects where there is duplication of effort. In the context of environmental review requirements, such duplication could occur if states were required to carry out two separate—but similar—analyses to satisfy federal and state requirements, for example, but not if the same analysis could be used to satisfy both state and federal documentary (i.e., procedural) requirements. In our interviews with state and FHWA officials, we inquired about duplication within and among highway projects, as well as duplication that may occur across time within a project. We also asked FHWA and state officials about the cost of any potential duplication (and how such cost might be measured) and the frequency of any potential duplication. Finally, we asked state officials about efforts to make the environmental review process more efficient in our survey, as well as the potential benefits of any such efforts.

Table 2. States Agencies in Selected States with State Environmental Policy Laws (SEPA) Interviewed by GAO

State	State Agencies Interviewed
California	California Department of Transportation (Caltrans) California State Association of Counties The Governor's Office of Planning and Research
Hawaii	Hawaii Department of Transportation
Maryland	Maryland Department of Transportation, State Highway Administration, Office of Planning and Preliminary Engineering, Environmental Planning Division Critical Area Commission for the Chesapeake and Atlantic Coastal Bays
Massachusetts	Massachusetts Executive Office of Energy and Environmental Affairs, Massachusetts Environmental Policy Act Massachusetts Department of Transportation, Highway Division
Minnesota	Minnesota Environmental Quality Board Minnesota Department of Transportation
North Carolina	North Carolina Department of Transportation North Carolina Department of Environment and Natural Resources
New York	New York State Department of Transportation
Washington	Washington State Department of Transportation
Wisconsin	Wisconsin Department of Transportation Wisconsin Department of Natural Resources

Source: GAO and state agencies. | GAO-15-71

We conducted this performance audit from May 2013 to November 2014 in accordance with generally accepted government auditing standards. Those standards require that we plan and perform the audit to obtain sufficient, appropriate evidence to provide a reasonable basis for our findings and conclusions based on our audit objectives. We believe that the evidence obtained provides a reasonable basis for our findings and conclusions based on our audit objectives.

End Notes

[1] Although most of the nation's roads are owned by state and local governments, the federal government supports state investment in highway construction via a grant-based cost reimbursement system. In 2012, the most recent year for which data are available, $38 billion was obligated to states. For purposes of this report, we are focusing on federal- aid highway projects (projects authorized under title 23 of the United States Code) such as projects involving highways or bridges.

[2] Pub. L. No. 91-190 (1970), codified at 42 U.S.C. §§ 4321-4347.

[3] NEPA states that the purposes of the act are "to declare a national policy which will encourage productive and enjoyable harmony between man and his environment; to promote efforts which will prevent or eliminate damage to the environment and biosphere and stimulate the health and welfare of man; to enrich the understanding of the ecological systems and natural resources important to the Nation; and to establish a Council on Environmental Quality." 42 U.S.C. § 4321.

[4] NEPA requirements apply to both major and minor projects. The level of review required for a project depends on the potential for environmental effects and several other factors, as discussed below.

[5] NEPA applies to federal agency policies, programs, plans, and projects. Projects include actions approved by permit or other regulatory decisions as well as federal and federally assisted activities, including projects or actions funded by or undertaken by FHWA. 40 C.F.R. § 1508.18(b). For more information about how NEPA reviews are carried out, see GAO, *FAA Airspace Redesign: An Analysis of the New York/New Jersey/Philadelphia Project,* GAO-08-786 (Washington, D.C.: July 31, 2008) and *Highway Infrastructure: Stakeholders' Views on Time to Conduct Environmental Reviews of Highway Projects,* GAO-03-534 (Washington, D.C.: May 23, 2003).

[6] See GAO, *National Environmental Policy Act: Little Information Exists on NEPA Analyses,* GAO-14-369 and GAO-14-370 (Washington, D.C.: April 15, 2014); GAO, *Federal-Aid Highways: Federal Requirements for Highways May Influence Funding Decisions and Create Challenges, but Benefits and Costs Are Not Tracked,* GAO-09-36 (Washington, D.C.: Dec 12, 2008); and GAO, *Highway Projects: Some Federal and State Practices to Expedite Completion Show Promise,* GAO-12-593 (Washington, D.C.: June 6, 2012).

[7] MAP-21 authorized surface transportation programs, including federal-aid highway programs, through fiscal year 2014. Pub. L. No. 112-141, 126 Stat. 405 (2012). Legislation was enacted in August 2014, extending highway and other surface transportation programs and funding through May 31, 2015, for programs authorized under MAP-21. Highway and Transportation Funding Act, Pub. L. No. 113-159, 128 Stat. 1839 (2014).

[8] Pub. L. No. 112-141, § 1322, 126 Stat. at 553.

[9] See GAO, *Government Efficiency and Effectiveness: Strategies for Reducing Fragmentation, Overlap, and Duplication and Achieving Cost Savings,* GAO-13-631T (Washington, D.C.:

May 16, 2013). We have applied this definition in a body of work examining potentially duplicative activities in the federal government. See, for example, GAO, *Opportunities to Reduce Potential Duplication in Government Programs, Save Tax Dollars, and Enhance Revenue*, GAO-11-318SP (Washington, D.C.: Mar. 1, 2011); *2012 Annual Report: Opportunities to Reduce Duplication, Overlap and Fragmentation, Achieve Savings, and Enhance Revenue*, GAO-12-342SP (Washington, D.C.: Feb. 28, 2012); and *2013 Annual Report: Actions Needed to Reduce Fragmentation, Overlap, and Duplication and Achieve Other Financial Benefits*, GAO-13-279SP (Washington, D.C.: Apr. 9, 2013).

[10] FHWA's *Environmental Review Toolkit for NEPA and Transportation Decisionmaking* (FHWA's NEPA Toolkit) was accessed in 2014 at: http://environment

[11] We included entities which are defined as "states" under Title 23 and which have SEPAs required to be applied to highway projects. Thus, we did not include the Nevada/California–Tahoe Region, which has SEPA requirements, because it is not defined as a state under Title 23. We included New Jersey, which has an executive order that serves as a SEPA, and was included on CEQ's list of states with SEPAs. We also included Texas, which has recently enacted a SEPA for transportation projects and which CEQ officials believed should be included. We excluded South Dakota because although its SEPA provides the option of preparing an environmental impact statement, there is no requirement to do so, and South Dakota DOT officials told us that they do not conduct environmental reviews under the state law. Some other states also have specialized environmental review requirements even in the absence of a SEPA.

[12] We did not examine all details of federal and state laws and practices, nor did we examine all details of other sources such as case law that could affect how laws are implemented at the state and federal levels.

[13] 42 U.S.C. § 4332(2)(C).

[14] All 18 states we contacted responded to our survey, although one state did not complete one portion of the questionnaire.

[15] See Federal Water Pollution Control Act, Act of June 30, 1948, ch. 758, § 404, 62 Stat. 1155, codified as amended at 33 U.S.C. § 1344 (Clean Water Act).

[16] See Endangered Species Act of 1973, Pub. L. No. 93-205, § 7, 87 Stat. 884 (1973), codified at 16 U.S.C. § 1536.

[17] 40 C.F.R. § 1506.2(c).

[18] FHWA administers the Federal-Aid Highway Program and distributes most funds to the states through annual apportionments established by statutory formulas. Once FHWA apportions these funds, the funds are available for obligation to fund state-selected projects, but these cannot proceed to construction until the NEPA process is completed. States are primarily responsible for selecting projects and administering federal-aid projects through their state DOTs, as well as for providing matching funds. While state DOTs are the recipients of federal funds and administer and oversee many of the projects funded by the federal-aid highway program, states sometimes allow local public agencies, such as towns, cities, and counties, to administer federally funded projects. In these instances, states continue to be responsible for ensuring that federal funds are spent in accordance with federal law.

[19] 23 C.F.R. § 771.111(f)(1).

[20] It is FHWA's policy that to the fullest extent possible, all environmental investigations, reviews, and consultations be coordinated as a single process, and compliance with all applicable environmental requirements be reflected in the environmental document. 23 C.F.R. §§ 771.105, 771.111(d).

[21] 23 U.S.C. § 139(l)(1). Critics of NEPA have stated that those who disapprove of a federal project will use NEPA as the basis for litigation to delay or halt that project. Others argue that litigation only results when agencies do not comply with NEPA's procedural requirements. See GAOand GAO-14-370.

[22] The statute permits states to fund assistance to affected federal or state agencies and Native American tribes participating in the environmental review process to support activities

expediting and improving transportation project delivery, among other things. 23 U.S.C. §139(j)(1).

[23] Social demands influencing a project may include its impacts on employment, schools, or recreational facilities and land uses that alter highway capacity needs. The project's environmental justice analysis must identify and address all reasonably foreseeable adverse social, economic, and environmental effects on minority and low-income populations. See Executive Order 12898, *Federal Actions to Address Environmental Justice in Minority Populations and Low-Income Populations*, 59 Fed. Reg. 7629 (Feb. 16, 1994); U.S. Department of Transportation (DOT) Order 5610.2, *Environmental Justice in Minority and Low-Income Populations*; and FHWA Order 6640.23A, *FHWA Actions to Address Environmental Justice in Minority Populations and Low-Income Populations*.

[24] Section106 of the National Historic Preservation Act, Pub. L. No. 89-665, § 106, 80 Stat. 915 (1966), codified as amended at 16 U.S.C. § 470f, requires federal agencies to take into account the effects of a government-assisted or licensed project on any historic site, building, structure, or other object that is listed on the National Register of Historic Places.

[25] Section 4(f) of the Department of Transportation Act of 1966, Pub. L. No. 89-670, § 4(f), 80 Stat. 931, the substance of which now appears at 23 U.S.C. § 138 and 49 U.S.C. § 303, protects parklands from adverse effects of federal-aid highway projects. In general, section 4(f) requires that the impact be avoided or rendered *de minimis*, or that the state must show that there is no feasible and prudent alternative. The requirement dates from enactment of section 4(f) of the Act establishing DOT and predates NEPA. Pursuant to 23 U.S.C. § 139, FHWA includes section 4(f) in its NEPA analysis as an environmental law with the same status as the Clean Air Act or Clean Water Act.

[26] 16 U.S.C. §§ 1271-1287.

[27] As applicable here, the terms "federal lead agency" and "lead agency" come from 23 U.S.C. § 139, which governs these cases. The state DOT or other applicable state agency may serve with U.S. DOT as a "joint lead agency."

[28] To determine "significance" as defined by NEPA, responsible officials must consider both the context and intensity of an action's impact. 40 C.F.R. § 1508.27. This means responsible officials must consider beneficial as well as adverse impacts, long- and short- term, and cumulative impacts, as well as the impact of external requirements such as, for title 23-authorized projects, the impact on certain types of public lands and historic sites (see 23 U.S.C. § 138 and 49 U.S.C. § 303) and compliance with title 23-specific processes (see 23 U.S.C. § 139).

[29] See GAO

[30] Some EISs are completed in less time. For example, in 2013, the median time to complete EISs started since the enactment of SAFETEA-LU, in 2005, was less than 4 years.

[31] FHWA, *Report to Congress on Costs Associated with the Environmental Process: Impacts of Federal Environmental Requirements on Federal-aid Highway Project Costs*. (Washington, D.C.: Oct. 25, 2006). FHWA does not include the time needed to approve project permits in its calculations of the median time needed to complete NEPA reviews.

[32] FHWA regulations identify two types of CEs: listed CEs, which meet CEQ and FHWA requirements by regulation, and documented CEs, which only may be designated as CEs after FHWA approval of project-specific documentation. In some cases, states may be able to document a CE with minimal additional paperwork, while other documented CEs—for more complex projects—may require formal submission of additional information from the state to allow the FHWA division office to conclude that the project will not cause a significant environmental impact.

[33] For more information about SEPAs, see e.g., Daniel R. Mandelker, et al., *NEPA Law and Litigation*, Thomson Reuters/West, Rel. 10, August 2012.

[34] Some projects may be categorically excluded, or they may fall below a SEPA threshold for environmental review.

[35] Federal permitting requirements can also trigger a NEPA review requirement for state- funded highway projects. For example, a highway project without federal-aid highway funding may still trigger a NEPA review if a federal permit is required from the U.S. Army Corps of Engineers (Corps) or the U.S. Coast Guard (Coast Guard) if it will affect a navigable waterway or wetland, among other resources. In these cases, the Corps or the Coast Guard, as applicable, could serve as the federal lead agency.

[36] Projects must comply with applicable substantive resource protection laws regardless of whether NEPA or a SEPA (if any) applies to the project.

[37]Section 7 of the Endangered Species Act mandates that federally authorized or funded actions must not jeopardize a threatened or endangered species, and it requires that FHWA or a state DOT consult with and submit biological assessments to the U.S. Fish and Wildlife Service or the National Marine Fisheries Service, as appropriate. Section 10 of the Act applies to a project without federal authorization or funding—including projects with only state funding—and it requires a project sponsor to apply for a permit and submit a habitat conservation plan, which may be more burdensome, according to state officials. For more information on Section 7, see GAO, *Endangered Species Act: The U.S. Fish and Wildlife Service Has Incomplete Information about Effects on Listed Species from Section 7 Consultations,* GAO-09-550 (Washington, D.C.: May 21, 2009).

[38] We have previously found that states sometimes find it advantageous to use nonfederal funds for projects to avoid the costs or delays that may arise when complying with federal requirements such as Section 4(f). See GAOand GAO, *Federal-Aid Highways: Federal Highway Administration Could Further Mitigate Locally Administered Project Risks,* GAO-14-113 (Washington, D.C.: January 16, 2014).

[39] In January 2014, however, pursuant to MAP-21, FHWA adopted a regulation allowing a project that receives less than $5,000,000 of federal funds to be treated as a CE.

[40] FHWA regulations cite examples of projects that normally require an EIS—a new controlled access freeway, a highway project of four or more lanes on a new location, and new construction or extension of a separate roadway for buses or high occupancy vehicles not located within an existing highway facility. 23 C.F.R. § 771.115(a).

[41] Under FHWA's NEPA process, lead agencies are responsible for compliance with the requirements of all applicable environmental laws, including state laws. 23 C.F.R. § 771.133.

[42] As mentioned above, a FONSI is a finding of no significant impact. A NEPA FONSI cannot be prepared without having completed an EA review, and the FONSI includes or provides a summary of the EA. 40 C.F.R. § 1508.13.

[43] See appendix I for further discussion of our legal analysis.

[44] The requirements identified in FHWA's NEPA Toolkit are not exhaustive of all of FHWA's NEPA requirements; rather, FHWA highlights key requirements under each of six NEPA principles, including purpose and need, which was discussed above. We did not include questions related to purpose and need in our survey of state DOTs.

[45] The survey asked states to categorize their requirements as "less stringent," "similar," or "more stringent" than FHWA's NEPA requirements. States could also select "not applicable." By comparing our legal analysis to the responses of states who most frequently indicated that FHWA's NEPA requirements were "not applicable" to their SEPA, we determined that a "not applicable" response indicated less stringent requirements than FHWA NEPA or no comparable requirements. Therefore, we combined responses from the "less stringent" and "not applicable" categories.

[46] Officials with the Virginia DOT did not respond to the portion of our survey comparing the state's and FHWA's NEPA requirements. In our legal analysis, we found that Virginia's requirements were not similar to CEQ's or FHWA's NEPA requirements for the majority (at least 9 of 12) of the key requirements we assessed, including consideration of alternatives, mitigation, and public involvement. This analysis did not include specific requirements within each of the NEPA principles from FHWA's NEPA Toolkit.

[47] Classification of requirements as "similar" or "generally similar" does not indicate that they are identical. Similar as the term is used in this report, conveys that two items share a general resemblance, not that they are alike in all respects. Thus, we characterized states' survey responses as being similar if 75 percent or more of the questions about state requirements under a NEPA principle were marked as "similar" or "more stringent" than FHWA's NEPA requirements. If 50 to 74 percent of the requirements were marked as "similar" or "more stringent," we characterized states' survey responses as somewhat similar. If 51 to 74 percent of the requirements were marked as "less stringent" or "not applicable," we characterized states' survey responses as somewhat less stringent. If 75 percent or more of the requirements were marked as "less stringent" or "not applicable," we characterized states' survey responses as less stringent. Additional information about our methodology can be found in appendix II.

[48] In addition, legal requirements in Connecticut call for more limited consideration of cumulative impacts than NEPA (considering only those effects caused by the lead agency or the proposed projects), but state DOT officials told us that in practice the state DOT exceeds legal requirements when preparing an environmental analysis.

[49] As mentioned above, significance is measured by both its depth and breadth. In measuring whether a state's law is similar to NEPA and its implementing regulations in defining significance, we looked in part to whether state law requires consideration of direct, indirect, and cumulative effects. Direct effects are caused by and occur at the same time and place as the proposed action. Indirect effects are the secondary consequences on local or regional social, economic or natural conditions or resources that could result from additional activities (associated investments and changed patterns of social and economic activities) induced or stimulated by the proposed action, both in the short-term and in the long-term. Cumulative impacts are the impacts on the human and physical environment that result from the incremental impact of the proposed action when added to other past, present or reasonably foreseeable future action. We further discuss how the depth and breadth of an assessment of significant impacts varies among states in appendix I.

[50] According to U.S. DOT, the principles of environmental justice are: (1) avoiding, minimizing, or mitigating disproportionately high and adverse human-health and environmental effects, including social and economic effects, on minority populations and low-income populations; (2) ensuring the full and fair participation by all potentially affected communities in the transportation decision-making process; and (3) preventing the denial of, reduction in, or significant delay in the receipt of benefits by minority and low-income populations.

[51] Under CEQ regulations, economic or social effects do not by themselves require preparation of an EIS. When an EIS is prepared and economic or social and natural or physical environmental effects are interrelated, then the EIS will discuss all of these effects on the human environment. 40 C.F.R. § 1508.14.

[52] The "no action" alternative provides a benchmark, enabling decision makers to compare the magnitude of environmental effects of the action alternatives.

[53] In general, while some states address alternatives directly, others address them through regulations governing documentation requirements. Further, state documentation requirements sometimes distinguish between accounting for alternatives that were considered and reporting alternatives that were not considered.

[54] The District of Columbia's EPA Guidelines note that as with the federal regulations for implementing NEPA, "reasonable" alternatives include those that are practical or feasible from the technical and economic standpoint and using common sense.

[55] Minnesota requires that the scoping document for an EIS-type assessment address one or more alternatives for a number of alternative types (i.e., modified designs and alternative technologies).

[56] More information about the consideration and selection of alternatives can be found in appendix I.

[57] FHWA policy provides that federal-aid highway decisions should be made "in the best overall public interest based upon a balanced consideration of the need for safe and efficient transportation; of the social, economic, and environmental impacts of the proposed transportation improvement; and of national, state, and local environmental protection goals." 23 C.F.R. § 771.105(b).

[58] CEQ regulations also require a brief discussion of reasons why some alternatives were eliminated from consideration.

[59] NEPA requires consideration of mitigation, and FHWA requires implementation of mitigation measures if doing so represents a reasonable public expenditure.

[60] Appendix I provides examples of how states address mitigation.

[61] HAW. ADMIN. CODE §§ 11-200-10(7), 11-200-17(b)(3), 11-200-17(m).

[62] As mentioned above, different sections of the Endangered Species Act may apply, depending on whether the project involves federal authorizations or funding.

[63] This requirement is focused on ensuring that the state's growth management process and Model Toxics Control Act are integrated into environmental reviews.

[64] For most projects, including projects requiring an EA or EIS, 23 U.S.C. § 128 requires that states afford an opportunity for a public hearing (with transcript) for federal-aid highway projects to consider, among other matters, environmental issues.

[65] In several states, hearings are only held at the request of the project sponsor or an association with a minimum number of members (e.g., 25) who will be affected by the project.

[66] Washington DOT's local project search allows the public to search for projects by location—either using an address or an interactive map—and provides information such as funding sources, as well project cost, status, and projected improvements, among other things.

[67] We have defined duplication as occurring when two or more agencies or programs are engaged in the same activities or provide the same services to the same beneficiaries. See GAO

[68] Officials from the District of Columbia reported duplication between state and federal requirements in their survey, but explained that the requirements overlap and that state requirements provide an exemption for projects that have completed and approved NEPA documents. As a result, there is no duplication of effort to satisfy state requirements, and we have included the District of Columbia among the states reporting no duplication.

[69] State adoption of federal documents or analyses is discussed in more detail in appendix I.

[70] See GAO, *National Environmental Policy Act: Little Information Exists on NEPA Analyses,* GAO-14-369 and GAO-14-370 (Washington, D.C.: Apr. 15, 2014), citing a 2003 study funded by FHWA. The FHWA study evaluated the performance of environmental "streamlining" and noted that NEPA cost data would be difficult to segregate for analysis. (See U.S. Department of Transportation, Federal Highway Administration, *Evaluating the Performance of Environmental Streamlining: Phase II* (Washington, D.C.: 2003).) We also reported that FHWA and its 52 division offices have traditionally used an internal data system to track EIS documents. At the time, FHWA officials told us that the agency is in the process of replacing the agency's legacy system with the new Project and Program Action Information (PAPAI) system, which went online in March 2013. PAPAI is capable of tracking information on EISs, EAs, and CEs, including project completion time frames, but its use is not mandatory, according to DOT officials.

[71] In all, six of the nine states where we conducted in-depth interviews reported using checklists for their EA-type reviews, a practice that could result in some level of federal– state duplication.

[72] Puerto Rico in its survey responses identified duplication, but did not provide an explanation or examples of the duplication. Because we did not interview state DOT officials, we were unable to determine the type or cause of the potential duplication.

[73] CEQ regulations require that an agency must prepare a supplemental EIS if (a) the agency makes substantial changes in the proposed action that are relevant to environmental concerns, or (b) there are significant new circumstances or information relevant to environmental concerns and bearing on the proposed action or its impacts.

[74] Through the NEPA Assignment and CE Assignment Memorandums of Understanding, Caltrans is now responsible for FHWA's responsibilities under NEPA and other federal environmental laws such as the Endangered Species Act and Section 106 of the National Historic Preservation Act. This assignment was carried out under 23 U.S.C. §§ 326 and 327.

[75] According to officials with the California State Association of Counties, revisions add about $2,000 to $5,000 per technical report and typically delay highway projects by 1 to 3 months. In California, approximately 40 percent of federal-aid highway funds are for projects sponsored by Caltrans, according to state officials, while the remainder of the funds is used for projects sponsored by counties or other non-state project sponsors. Under the state's assumption of FHWA oversight responsibilities, Caltrans reviews and approves FHWA NEPA reviews for all projects receiving federal-aid highway funds.

[76] While states may approve categorical exclusions or make other decisions on FHWA's behalf under these programmatic agreements, FHWA remains legally subject to challenge for these decisions.

[77] According to FHWA, programmatic agreements are documents that establish a streamlined process for handling routine environmental requirements for commonly encountered project types. These agreements should clearly specify roles and responsibilities between state DOTs and other resource and regulatory agencies for consultation, review and compliance with one or more federal laws concerning cultural and historic preservation, environmental review processes and natural resource protection and conservation. The agreements usually set procedures for consultation, review and compliance with one or more federal laws.

[78] The first round of Every Day Counts was launched in 2010. It was expanded in 2012 to include 13 other innovations to build higher-quality and longer-lasting roads. FHWA announced a third round of Every Day Counts initiatives in 2014.

[79] Several turtles species are endangered in the state, including the plymouth red-bellied turtle entire (*Pseudemys rubriventris bangsi*), the sea turtle, hawksbill entire (*Eretmochelys imbricata*), the sea turtle, kemp's ridley entire (*Lepidochelys kempii*), and the sea turtle, leatherback entire (*Dermochelys coriacea*).

[80] Under the Texas SEPA, for example, the Texas Department of Transportation must "focus on delivering safe, efficient transportation projects and making sound decisions based on a balanced consideration of transportation needs and of social, economic and environmental impacts of proposed transportation improvements." 43 TEXAS ADMIN. CODE § 2.2.

[81] *See, e.g.,* MINN. STAT. § 116D.01(a); Mont. Code Ann. § 75-1-102(2); N.Y. ENVTL. CONSERV. LAW § 8-0101; WASH. REV. CODE § 43.21C.010.

[82] SEPAs in Massachusetts, New Jersey, New York, and North Carolina reflect this policy. *See* MASS. GEN. L^AWS, ch. 30, § 61; N.J. GOV. KEAN, EXEC. ORDER NO. 215; N.Y. ENVTL. CONSERV. LAW § 8-0109; N.C. Gen. Stat. § 113A-4.

[83] Several states limit their SEPAs to legislative, appropriations or planning functions. *See, e.g.,* M^D. CODE ANN., NAT. RES. § 1-301(d).

[84] In January 2014, pursuant to MAP-21, FHWA adopted a regulation allowing, among other things, projects that receive less than $5,000,000 of federal funds to be treated as a CE. 79 Fed. Reg. 2107, 2118-2119 (Jan. 13, 2014).

[85] 20 D.C. MUN. REGS. § 7201.1.

[86] GA. CODE ANN. § 12-16-3(7)(B).

[87] N.J. GOV. KEAN, EXEC. ORDER NO. 215.

[88] Texas excludes transportation projects that are not on the state highway system and are funded solely with special funds or that are being developed by a local governmental entity with no state or federal funds. 43 TEXAS ADMIN. CODE § 2.3(b)(1).

[89] CEQ's regulations broadly define "human environment" to include the natural and physical environment and the relationship of people with that environment. Economic and social effects are not intended by themselves to require preparation of an EIS, but when an EIS is prepared and economic or social and natural or physical environmental effects are

interrelated, the EIS must discuss all of these effects on the human environment. 40 C.F.R. § 1508.14.

[90] The types of analysis required are an EIS, EA, CE, and an environmental report, a brief document provided for under WisDOT regulations and used by WisDOT to demonstrate that a proposed action meets the criteria or conditions for a categorical exclusion. *See* WIS. ADMIN. CODE TRANS. §§ 400.04(9) (EIS), 400.07(2)(b) (EA), 400.04(3) (CE), 400.04(10) (environmental report), 400.08 (categorization of transportation projects).

[91] Texas has applied for participation in the Surface Transportation Project Delivery Program, which allows for states to apply to assume, and for FHWA to assign, environmental review responsibilities under NEPA and part or all of FHWA's responsibilities for environmental reviews, consultations, or other actions required under any federal environmental law with respect to one or more federal highway projects within the state. 79 Fed. Reg. 61370 (Oct. 10, 2014). FHWA has determined that the application is complete and has developed a draft Memorandum of Understanding with Texas, available for public comment, outlining how the state will implement the program with FHWA oversight. *Id.*

[92] *See* Memorandum of Agreement for the Review of Highway Projects Undertaken by the Va. Dep't of Transp. (Nov. 27, 2001).

[93] Section 4(f) refers to the Department of Transportation Act, Pub. L. No. 89-670, § 4(f), 80 Stat. 931, 934 (1966), which restricted use of any land in a public park, recreation area, wildlife or waterfowl refuge, or historic site unless there was no feasible alternative. That statute and others were recodified and the substance of the "section 4(f)" parkland requirements are now contained in 23 U.S.C. § 138 and 49 U.S.C. § 303.

[94] 20 D.C. MUN. REGS. 7299.1.

[95] Ga. CODE ANN. § 12-16-8 (3).

[96] HAW. CODE R. § 11-200-12.

[97] 326 IND. ADMIN. CODE § 16-2.1-6(5).

[98] MD. CODE REGS. 11.01.08.03(B)(8)(c), 11.01.08.08.

[99] N.J. GOV. KEAN, EXEC. ORDER NO. 215.

[100] VA. ANNO. CODE §§ 10.1-1188 to 10.1-1192.

[101] WIS. ADMIN. CODE TRANS. §§ 400.08(1)(d), 400.10(4)(7).

[102] *See, e.g.,* MASS. GEN. LAWS, ch. 30, § 62B (requiring consideration of "reasonable alternatives to the proposed project and their environmental consequences"); WASH. ADMIN. CODE § 197-11-440(5) (reasonable alternatives must be considered); N.Y. COMP. CODES R. & REGS., tit. 6, § 617.9(b)(5)(v) (must consider a "range of reasonable alternatives to the action that are feasible, considering the objectives and capabilities of the project sponsor"); MONT. CODE ANN. § 75-1-201(1)(b)(iv)(C) ("Alternatives considered must be reasonable" and only those using current technology and which are economically feasible must be included).

[103] *See* MASS. GEN. LAWS, ch. 30, § 61 (state "shall use all practicable means and measures to minimize damage to the environment"); N.Y. ENVTL. CONSERV. § 8-0109.2(f); N.Y. COMP. CODES R. & REGS., tit. 6, § 617.11(d)(5) (final EIS must "certify that consistent with social, economic and other essential considerations from among the reasonable alternatives available, the action is one that avoids or minimizes adverse environmental impacts to the maximum extent practicable.").

[104] *See* 40 C.F.R. §§ 1500.2(c) (integration of environmental review and planning procedures), 1501.5 (selection of lead agencies), 1501.6 (designation of cooperating agencies),1500.4(i) (use of tiering to reduce repetitive analyses; tiering means use of a broad EIS to frame multiple subsequent EISs; *see* 43 Fed. Reg. 55978, 55979 (Nov. 29, 1978)), 1500.4(n) (joint preparation and adoption of environmental analyses to reduce duplication), 1500.5(h) (same, to reduce delay),1502.5(b) (encouraging federal agencies to begin preparation of environmental assessments or statements early, preferably jointly with applicable state or local agencies), 1506.2 (requiring cooperation to eliminate duplication with state and local procedures). For FHWA regulations, *see* 23 C.F.R. part 771.

[105] This requirement for collaboration is viewed as enhancing the overall efficiency of the NEPA process as applied to federal-aid highway projects and is also reflected in the requirements for coordination and for a "single-process review," discussed in Elements 8 and 9 below, as well as in authorization by many SEPA states of the state's use of some or all of a project's NEPA documentation or analysis, discussed in Element 10 below.

[106] 23 U.S.C. § 139(b)(3)(A).

[107] Growth Management Act, WASH. REV. CODE Ch. 36.70A; Model Toxics Control Act, WASH. REV. CODE Ch. 70.105D.

[108] This is authorized at the federal level under CEQ's regulations. *See* 40 C.F.R. §§ 1500.4(n), 1506.2, 1506.3 (eliminating duplication with State and local procedures, by providing for joint preparation, and with other federal procedures, by providing that a federal agency may adopt appropriate environmental documents prepared by another agency).

[109] The APA allows aggrieved parties to challenge "final agency action for which there is no other adequate remedy in a court," 5 U.S.C. § 704, which includes FHWA issuance of a NEPA Record of Decision. *See, e.g., Sierra Club v. Slater,* 120 F.3d 623 (6th Cir. 1997).

[110] FHWA division offices, in consultation with state DOTs, can decide whether to publish a notice in the *Federal Register*. In the absence of a notice, a claim for judicial review must be filed within 6 years of the agency decision. *See* 28 U.S.C. § 2401(a) (general six-year statute of limitations for civil actions against the United States).

[111] Model State Admin. P. Act (1961), § 15; Model State Admin. P. Act (1981), §§ 5-101 to 5-205.

[112] Pub. L. No. 112-141, § 1322, 126 Stat. 405, 552 (2012).

[113] Pub. L. No. 91-190 (1970), codified at 42 U.S.C. §§ 4321-4347.

[114] See GAO, *National Environmental Policy Act: Little Information Exists on NEPA Analyses,* GAO-14-369 (Washington, D.C.: Apr. 14, 2014); GAO, *Federal-Aid Highways: Federal Requirements for Highways May Influence Funding Decisions and Create Challenges, but Benefits and Costs Are Not Tracked,* GAO-09-36 (Washington, D.C.: Dec. 12, 2008); and GAO, *Highway Projects: Some Federal and State Practices to Expedite Completion Show Promise,* GAO-12-593 (Washington, D.C.: June 6, 2012).

[115] 42 U.S.C. § 4332(2)(C).

[116] One state did not complete the portion of the survey comparing state and federal requirements.

[117] The six principles were derived from FHWA's *Environmental Review Toolkit for NEPA and Transportation Decision Making,* which was accessed in 2014 at: http://environmentThe requirements identified in FHWA's NEPA Toolkit are not exhaustive of all FHWA's NEPA requirements; rather, FHWA highlights key requirements under each of six areas identified as a NEPA principle, and discusses documentation and disclosure. We asked states to compare their requirements with FHWA requirements for five of these principles in our survey, among other things. A sixth principle—establishing the purpose and need of a project—is essential in establishing a baseline for the development of the range of reasonable project alternatives required in environmental reviews. In the survey, we asked state DOT officials to report on the whether their state requires documentation of their discussion about the purpose and need of a proposed project for EIS-type reviews, but we did not ask them to compare specific requirements associated with establishing the purpose and need for a project.

[118] Part IV of the survey asked states to categorize their requirements as "less stringent," "similar," or "more stringent" than FHWA's NEPA requirements. States could also select "not applicable." By comparing our legal analysis to the responses of states who most frequently indicated that FHWA's NEPA requirements were "not applicable" to their SEPA, we determined that a "not applicable" response indicated less stringent requirements than FHWA NEPA or no comparable requirements. Therefore, we combined responses from the "less stringent" and "not applicable" categories.

[119] We surveyed South Dakota DOT officials before determining that the state does not have a SEPA that is mandatory for transportation projects. We excluded the state from our analysis.

[120] For the purposes of reporting the survey results at a high level, we determined that a "more stringent" response is like a "similar" response in that the state requirement is at least comparable to FHWA's NEPA requirements. In addition, for the 42 individual requirements about which we surveyed, no state reported more than 4 state requirements as more stringent than FHWA's NEPA requirements. Therefore, for this analysis we combined responses from the "similar" and "more stringent" categories.

[121] See GAO, *Government Efficiency and Effectiveness: Strategies for Reducing Fragmentation, Overlap, and Duplication and Achieving Cost Savings,* GAO-13-631T (Washington, D.C.: May 16, 2013). GAO has applied this definition in numerous reports examining potentially duplicative activities in the federal government. See also GAO, *Opportunities to Reduce Potential Duplication in Government Programs, Save Tax Dollars, and Enhance Revenue,* GAO-11-318SP (Washington, D.C.: Mar. 1, 2011); *2012 Annual Report: Opportunities to Reduce Duplication, Overlap and Fragmentation, Achieve Savings, and Enhance Revenue,* GAO-12-342SP (Washington, D.C.: Feb. 28, 2012); and *2013 Annual Report: Actions Needed to Reduce Fragmentation, Overlap, and Duplication and Achieve Other Financial Benefits*, GAO-13-279SP (Washington, D.C.: Apr. 9, 2013).

In: Federal and State Environmental Reviews … ISBN: 978-1-63482-390-6
Editor: Marlene Wilkinson

Chapter 2

THE ROLE OF THE ENVIRONMENTAL REVIEW PROCESS IN FEDERALLY FUNDED HIGHWAY PROJECTS: BACKGROUND AND ISSUES FOR CONGRESS[*]

Linda Luther

SUMMARY

Under programs administered by the Department of Transportation's (DOT's) Federal Highway Administration (FHWA), certain highway and bridge projects may be eligible for federal funding. Project approval and the receipt of federal funds are conditioned on the project sponsor (e.g., a local public works or state transportation agency) meeting certain standards and complying with federal law. Activities necessary to demonstrate compliance with those requirements may be completed at various stages of project development. Although the names of each stage may vary from state to state, project development generally includes the following: planning, preliminary design and environmental review, final design and rights-of-way acquisition, construction, and facility operation and maintenance.

[*] This is an edited, reformatted and augmented version of the Congressional Research Service Publication, CRS Report for Congress R42479, dated April 11, 2012.

When there is debate over the time it takes to complete federal highway projects, the environmental review stage has been a primary focus of congressional attention concerning legislative options to speed project delivery. The current process includes activities necessary to demonstrate that all potential project-related impacts to the human, natural, and cultural environment are identified; effects of those impacts are taken into consideration (among other factors such as economic or community benefits) before a final decision is made; the public is included in that decision-making process; and all state, tribal, or federal compliance requirements applicable as a result of the project's environmental impacts are, or will be, met.

Compliance requirements depend on site-specific factors, including the size and scope of the project, and whether and to what degree it may affect resources such as parks, historic sites, water resources, wetlands, or urban communities. For all proposed federal-aid highway projects, however, some level of review will be required under the National Environmental Policy Act of 1969 (NEPA, 42 U.S.C. §4321 et seq.). Broadly, NEPA requires federal agencies to consider the environmental effects of an action before proceeding with it and to involve the public in the decision-making process.

The time it takes to complete the NEPA process is often the focus of debate over project delays attributable to the overall environmental review stage. However, the majority of FHWA-approved projects require limited documentation or analyses under NEPA. Further, when environmental requirements have caused project delays, requirements established under laws other than NEPA have generally been the source. This calls into question the degree to which the NEPA compliance process is a significant source of delay in completing either the environmental review process or overall project delivery. Causes of delay that have been identified are more often tied to local/state and project-specific factors, primarily local/state agency priorities, project funding levels, local opposition to a project, project complexity, or late changes in project scope. Further, approaches that have been found to expedite environmental reviews involve procedures that local and state transportation agencies may implement currently, such as efficient coordination of interagency involvement; early and continued involvement with stakeholders interested in the project; and identifying environmental issues and requirements early in project development.

Bills in the House and Senate (the American Energy and Infrastructure Jobs Act of 2012 (H.R. 7) and Moving Ahead for Progress in the 21st Century (MAP-21; S. 1813)) would reauthorize DOT programs. Both include provisions intended to expedite project delivery by changing elements of the environmental review process, particularly NEPA requirements. This report provides information on existing NEPA and environmental review requirements, particularly requirements that may be subject to change under the House and Senate proposals.

INTRODUCTION

Under programs administered by the Department of Transportation's (DOT's) Federal Highway Administration (FHWA), certain highway and bridge projects may be eligible to receive federal-aid funding.[1] As a condition of receiving those funds, a project sponsor (e.g., a local or state transportation agency) must meet certain standards and requirements applicable to activities completed at every stage of project development. Although the names of those stages may vary somewhat from state to state, those stages generally include initial project planning, preliminary design/engineering and environmental review, final design and rights-of-way acquisition, construction, and facility operation and maintenance.

Each stage of project development is initiated and completed largely at the state or local level, with FHWA having ultimate responsibility for ensuring that individual projects comply with requirements applicable to federal-aid highways.[2] Also, each development stage involves a range of activities that will affect the time it takes to deliver the project. Required elements of the preliminary design and environmental review stage will vary by project, but generally include processes necessary to identify and demonstrate compliance with environmental requirements applicable to that project.

When there is debate over the time it takes to complete federally funded highway projects,[3] particularly debate over activities that may expedite or delay project delivery, various elements of the environmental review stage of project development have been the focus of attention. However, whether or the degree to which elements of that process may delay projects is unclear.[4]

The two most recent laws authorizing DOT programs included requirements intended to expedite the environmental review process that focused primarily on procedures necessary to demonstrate compliance with the National Environmental Policy Act of 1969 (NEPA, 42 U.S.C. §4321 et seq.).[5] Current legislation to authorize DOT programs in the House and the Senate (the American Energy and Infrastructure Jobs Act of 2012 (H.R. 7) and Moving Ahead for Progress in the 21st Century (MAP-21; S. 1813)) also include provisions intended to expedite project delivery that focus primarily on the NEPA process.[6]

Despite the focus on the NEPA process, it is unclear whether or how changes to that process would result in faster highway project delivery. Available evidence regarding potential causes of project delays associated with environmental compliance is largely anecdotal and specific to unique, individual projects. Still, that evidence, while limited, points to issues or requirements apart from NEPA as more common causes of project delays.

This report identifies issues relevant to the debate over the role of the environmental review process in transportation project delivery. It identifies social and environmental issues that led Congress to enact the range of requirements that now make up the environmental review process, as well as selected requirements applicable to its implementation (particularly NEPA requirements). The report also identifies complexities in tying the environmental review process to federal-aid highway project delivery time. In particular, it identifies issues that make it difficult to determine the time it takes to complete the project development process, in general, or individual stages of development (e.g., activities related explicitly to environmental reviews); or to identify root causes of project delays tied to specific elements of the environmental review process. This report also discusses various approaches identified by transportation stakeholders as those that have expedited the environmental review process and overall project delivery.

Information and issues in this report were selected to help Members of Congress and their staff understand the NEPA compliance process as well as additional environmental compliance requirements that may be affected by H.R. 7 and MAP-21.

Background and Overview of Issues

Activities that may take place during the environmental review process and how that process is implemented will vary from project to project, from state to state. The environmental review process does not involve compliance with a single federal compliance requirement. It involves processes necessary to demonstrate compliance with a potentially wide array of requirements applicable to projects approved under the Federal-aid Highways program. Broadly, for federally funded highway projects, it involves two separate, but related processes—preparing appropriate documentation required under

NEPA; and identifying and demonstrating compliance with any additional state, tribal, or federal environmental requirements applicable to that project.

For a given project, how NEPA and other environmental compliance requirements must be demonstrated will largely depend on the degree to which the proposed project would have adverse effects on communities, natural or cultural resources (e.g., wetlands, endangered species habitat, historic sites, parks, or recreation areas), or special status land (e.g., farmland, floodplains, or coastal zones). Compliance with those requirements may include obtaining a permit, approval, study, or some level of analysis or consultation from an agency outside DOT.

NEPA was intended, in part, to ensure that federal agencies would consider the environmental impacts of an action among other factors (e.g., economic or community benefits) in the federal decision-making process. NEPA has two primary aims—to assure that federal agencies consider the environmental effects of their actions *before* proceeding with them and to involve the public in the decision-making process.

NEPA does not require an agency to elevate environmental concerns above other factors in the overall federal decision-making process. If the adverse environmental effects of a proposed action are adequately identified and evaluated, an agency is not constrained by NEPA from deciding that other project benefits outweigh the environmental costs and moving forward with the action. In contrast, other requirements applicable to federal-aid highways *may* dictate or somehow affect the outcome of a project decision. For example, other federal laws may require the selected project alternative to be the one with the least impact to a particular resource, prohibit FHWA approval of a project alternative that uses certain resources, require certain mitigation measures to limit a project's impacts, or require that certain activities take place in accordance with certain criteria (e.g., as specified in a permit or approval).

Environmental Reviews and Project Delays

Required elements of the environmental review process, particularly compliance with NEPA, will have an effect on project development. For example, before DOT can approve a project and allow final project design, property acquisition, or project construction to proceed, the project sponsor must appropriately document compliance with NEPA and complete any investigation, review, or consultation necessary to demonstrate compliance with other applicable environmental requirements. Further, it is DOT policy to use the NEPA compliance process as a mechanism to balance transportation decision making by taking into account the potential impacts on the human and natural environment and the public's need for safe and efficient transportation.[7]

State and local transportation agency officials and other stakeholders with an interest in transportation improvement generally acknowledge that elements of the environmental review process provide important protections to the human, cultural, and natural environment. However, those officials also sometimes argue that completing the process can be difficult and time-consuming. Some have argued, for example, that the time it takes to complete required NEPA documentation and supporting analysis or to obtain required input or approval from outside agencies can delay completion of federally funded transportation projects.

It is generally not disputed that the time it takes to complete the environmental review process for federally funded highway projects can take months or even years. What is unclear is the degree to which elements of the environmental review process directly or routinely delay project delivery. Determining the time it takes to complete activities associated with the environmental review process, or delays directly attributable to those activities, is difficult for several reasons including, but not limited to:

- **Limits to available data.** There is no centralized source of data regarding highway project delivery. States generally do not track project development time from planning to construction. States generally do not attempt to isolate elements of the environmental review process, which may overlap with preliminary project planning, design, or engineering activities. Further, there is no standard measure for determining when a project or the environmental review process, in particular, is completed "quickly" or would be considered "delayed."
- **The influence of local factors on project delivery.** The environmental review process may start, stop, and restart for reasons unrelated to environmental issues. Local and state issues have been shown to have the most significant influence on whether a project moves forward relatively quickly or takes longer than anticipated. Those issues include the project's level of priority among others proposed in the state; changes in funding availability; and local

controversy or opposition to the project (which may or may not be connected to environmental issues).

- **The variation in project type and complexity.** The wide range of projects approved under programs administered by FHWA (e.g., bridge repair versus major new highway construction) do not easily allow an "apples to apples" comparison of the time it takes to complete the environmental review process or factors that may delay it. Anecdotal evidence regarding projects identified as "delayed" have involved multiple, complex causes of delay (including local issues) unique to *that* project, not a single cause that may be commonly applicable to other projects.
- **Variation among state requirements and implementation processes**. The effect of requirements under federal law may be difficult to isolate since local, state, or tribal requirements and procedures will also affect how environmental compliance requirements are implemented. State DOTs implement their project delivery process differently, depending on factors specific to their state and its needs. For example, some states may implement unique design and contracting processes that expedite project delivery that other states do not.
- **Time "saved" cannot be gauged.** Depending on the scope and complexity of the project, more time spent addressing environmental issues in the project planning and preliminary design stage may result in faster completion of final design and project construction (when delays may require actions that take more time and money to address). Time may also be saved when adverse project impacts that could lead to local opposition to the project are identified and addressed during the early stages of project development.

Challenges to Tying Project Delays to NEPA Compliance

Transportation agency officials and project sponsors have broadly identified environmental compliance requirements as a common source of frustration in completing the project development process. However, limits to and contradiction in available data make it difficult to clearly identify specific causes of delay that are directly and routinely attributable to specific elements of environmental compliance. Identifying a distinct root cause of a delay will arguably be necessary before effective "solutions" (procedures that would result in faster project delivery) can be identified. That is, knowing *that* a delay occurred may be irrelevant if it is not determined *why* the delay occurred. An understanding of *why* is useful in identifying a solution that directly addresses a problem's underlying cause.

Determining why a project was delayed may be difficult or may be attributable to multiple, interrelated factors. Generally, the more complex the project, the more complex the potential cause(s) of delay. For example, compared to a maintenance or repair project, a major new construction project will require more extensive review, documentation, or analysis to demonstrate compliance with NEPA and other applicable environmental requirements. However, the following factors call into question the degree to which NEPA alone is a significant source of project delay in overall project development:

- **The majority of projects require limited review under NEPA.** The majority of FHWA-approved projects (approximately 96%) involve no significant environmental impacts and, hence, require limited documentation, analysis, or review under NEPA.
- **Compliance with DOT's "NEPA regulations" extends beyond what is required under NEPA.** DOT's "Environmental Impact and Related Procedures"[8] prescribe the policies and procedures to ensure that FHWA-approved projects will comply with NEPA as well as requirements established under Title 23 applicable to Federal-aid Highways (e.g., provisions applicable to the consideration of adverse economic, social, and environmental effects (under §109(h)), public hearings (§128), and preservation of parklands (§138)).
- **The NEPA compliance process is used to demonstrate compliance with all applicable environmental review requirements.** It is DOT policy that any investigation, review, or consultation necessary to demonstrate compliance withapplicable environmental requirements be completed within the context of the NEPA process. This use of NEPA as an "umbrella" compliance process can blurthe distinction between what is required under NEPA and what is required under separate authority.

Transportation agency officials asked to identify sources of frustration or delay in completing the environmental review process most commonly cite compliance requirements applicable to the protection of parklands, historic sites, wetlands, or threatened or endangered species. The potential root cause of delay in complying with those requirements could be attributable to a wide range of project-specific factors (e.g., incomplete permit applications, challenges in obtaining multiple approvals or permits for a complex project, or disagreement with a resource agency over appropriate methods to mitigate project impacts).

Both existing law and regulations implementing NEPA include explicit directives and requirements intended to streamline the NEPA process. Included among those requirements are procedures intended to coordinate efficient agency interaction and cooperation, reduce NEPA-related paperwork and duplication of effort (e.g., documentation and analysis that may be required by similar state, tribal, or federal requirements or from one stage of project development to the next), and integrate the consideration of environmental compliance issues in a project's planning stage. Barriers to efficiently implementing existing requirements may be project-specific or involve issues that may be difficult to address by simply amending or eliminating existing federal requirements.

This is not to suggest that there are not instances where preparation of documentation and analysis required under NEPA is not time-consuming or may contribute to delays in project delivery. However, it is unclear whether or what additional federal requirements may be implemented to expedite the NEPA process. Conversely, it is not clear whether the elimination of certain NEPA-specific requirements may expedite project delivery or would alter the framework for coordinating an already complex compliance process, resulting in additional project delay. For a given project, whether changes to the NEPA process might result in faster project delivery will likely depend on the project's scope and complexity; the degree to which it is affected by "local" factors (e.g., state funding or project priorities); and compliance requirements applicable to the project, in addition to those under NEPA.

Highway Construction Impacts That Led to the Current Process

To understand why a complex array of requirements may apply to highway projects, it is useful to understand the social and environmental concerns that led Congress to enact the various laws that now form the framework of the environmental review process. Each requirement included within that process represents past efforts by Congress to minimize adverse impacts from federally funded highway projects or to minimize adverse impacts to certain communities or resources that Congress identified as needing some level of protection.

The current debate over the environmental review process frequently centers around the effect that completion of that process has on project delivery. The debate rarely recognizes the issues that led Congress to enact the requirements that now make up that process. Requirements included within the environmental review process, and procedures to demonstrate compliance with them, have evolved over the past 50 years. However, many of the requirements that are subject to particular scrutiny today were enacted between 1966 and 1972.

During the 1950s and 1960s, the public was becoming increasingly aware of and concerned about the impacts that human activity were having on the environment. Increasing attention turned to the effect that federally funded programs and projects were having on the human, cultural, and natural environment. One federal program that generated particular concern was the development and construction of the Interstate Highway System.

The Federal-Aid Highway Act of 1956 (P.L. 84-627) authorized and provided revenue sources for the construction of the National System of Interstate and Defense Highways (commonly known as the Interstate Highway System, Interstate System, or the Interstate). The Interstate System is a network of limited-access roads including freeways, highways, and expressways forming part of the National Highway System of the United States.[9] Construction of the Interstate System took approximately 35 years and resulted in a network of roads and bridges that currently includes over 45,000 miles of rural highways, suburban and urban freeways, and bridges.[10]

Although the connection of rural, urban, and suburban communities resulted in a host of economic and cultural benefits, construction of the Interstate System also brought certain adverse impacts to both the human and natural environment. Those impacts were seen particularly in the construction of the urban freeways. Planning for such projects often involved locating freeways within available open space or where land acquisition costs were relatively low. "Available open space" often meant historic sites, parks, or recreation areas. Adverse impacts to those resources from highway projects drew increased attention from newly formed stakeholder groups with an interest in environmental protection and historic preservation.

Project planning that involved lower land acquisition costs often meant property acquisitions in densely populated, working-class or high-poverty neighborhoods. Resulting urban freeway projects had a disproportionate impact on the urban poor. One such example involved a segment of I-95 north of Miami. The route selected by local transportation officials cut through the inner-city community of Overtown, a once-thriving African-American community known as the "Harlem of the South." A 2009 FHWA report discussing lessons learned in complying with environmental requirements describes the project as follows:

> In 1957, the Overtown community was almost decimated by the development of the I-95 and I-395 freeways. The constructed roadway had a disastrous impact on the economic and social structure of the community. The community continues to shoulder the lingering effects of those negative impacts, and as a result there is also persistent anger towards and distrust of [the Florida Department of Transportation].[11]

Opposition to other urban freeway projects led to "freeway revolts" spearheaded by newly established environmental and social justice groups.[12] Freeway revolts took place in cities like Baltimore, Boston, Los Angeles, New Orleans, New York, Reno, and San Francisco. As a result, a significant number of projects were abandoned or significantly scaled back due to widespread public opposition, especially by those whose neighborhoods would be disrupted or who would displaced by the proposed freeways.

ELEMENTS OF THE ENVIRONMENTAL REVIEW PROCESS

By the mid to late 1960s, Congress began to enact legislation intended to address the growing public concern over projects implemented under the Federal-aid Highways program. During that period, Congress also enacted legislation in response to increasing awareness and concern over the impacts of all federal actions—not just federal highway projects. Also during the 1960s and into the 1970s Congress began to enact a wide range of laws intended to identify, prohibit, control, or mitigate adverse impacts of human activities to specific community, natural, or cultural resources that Congress identified as in need of certain protection. This report identifies and summarizes requirements that have been identified as those most commonly applicable to federally funded highway projects.

Requirements Applicable to Federal-aid Highways

FHWA is prohibited from approving a project for funding under the Federal-aid Highway program until the project sponsor demonstrates that the proposed project will comply with all applicable federal, tribal, and state requirements. To the extent possible, compliance with any requirements that apply to a project, as a result of that project's effect on the human and natural environment, must be appropriately documented and demonstrated during the environmental review stage of project development.

Requirements specific to Federal-aid Highways include a host of standards, procedures, and conditions applicable to the various stages of project development. Several requirements (applicable primarily to activities that take place during the project planning, preliminary design, and environmental review phases of development) reflect concern over the effects of urban freeway construction (discussed above), including the following:

- **Directive to establish guidelines to assure consideration of adverse project impacts (23 U.S.C. §109(h)).** Directed DOT to establish guidelines to assure that possible adverse, economic, social, and environmental effects of proposed highway projects and project locations were fully considered during project development, and that final project decisions be made in the best overall public interest, taking into consideration the costs of eliminating or minimizing adverse effects to air, noise, and water pollution; destruction or disruption of man-made and natural resources; aesthetic values, community cohesion, and the availability of public facilities and services; adverse employment effects, and tax and property value losses; and injurious displacement of people, businesses, and farms.
- **Directive to establish noise standards (23 U.S.C. §109(i)).** Directed DOT to establish standards for highway noise levels compatible with different land uses. DOT cannot approve plans and specifications for any proposed federal-aid project unless it includes adequate measures to implement those noise standards. As implemented under DOT regulations, a project may be required to demonstrate compliance with applicable standards through an analysis of traffic noise impacts and, when necessary, to implement noise abatement measures.
- **Public hearings requirements (23 U.S.C. §128).** For a proposed project bypassing or going through any city, town, or village, a state transportation department is required to certify that it held or afforded

the opportunity for public hearings; considered the economic and social effects of the project location, and its impact on the environment; and considered the consistency of the project with local planning goals and objectives.

- **Preservation of parklands requirements (23 U.S.C. §138).** More commonly referred to as "Section 4(f)"[13] requirements, DOT is prohibited from approving a project that uses publicly owned (local, state, or federal) parks and recreation areas, wildlife and waterfowl refuges, and publicly or privately owned historic sites of national, state, or local significance. DOT may approve a project that uses a 4(f) resource only if there is no prudent and feasible alternative to do otherwise, and that use includes all possible planning to minimize harm to the resource.

Of the requirements specifically applicable to Federal-aid Highways, the preservation of parklands requirements may have the greatest effect on highway project development and delivery. Projects that would use a 4(f) resource require an evaluation analyzing project alternatives (including location and design shifts) that avoid the resource.[14] To be approved by FHWA, the evaluation must show that alternatives that would not use the resource would result in "truly unique problems," resulting in costs or community disruption of extraordinary magnitude. This test is often referred to as the "Overton Park Criteria," after a court case in the 1970s in Memphis, TN.[15] In approving the use of a 4(f) resource, FHWA must also consider the significance and importance of the resource itself.

SAFETEA amended Section 138 to allow for the use of a 4(f) resource if that use can be proven to have *de minimis* impacts to the resource.[16] Generally, *de minimis* impacts would result from the use of minor amounts of a particular resource. Such a determination requires concurrence from an official with jurisdiction over the resource. For example, for a transportation project adjacent to a publicly owned park, recreation area, or wildlife and waterfowl refuge, FHWA would be required to consult with, as appropriate, agencies within the Department of the Interior (e.g., the U.S. Fish and Wildlife Service, National Park Service, or the Bureau of Indian Affairs) or state or local park authorities. For historic sites, a *de minimis* impacts determination must be based on criteria established under the National Historic Preservation Act applicable to uses that will have no "adverse effect" on the site (16 U.S.C. §470f). The determination must receive concurrence from the State Historic

Preservation Officer (SHPO) and, if appropriate, the Advisory Council on Historic Preservation (ACHP).

Compared to other environmental requirements likely applicable to federal-aid highway projects, Section 4(f) is unique in its limits on the use of a protected resource. Most requirements intended to protect communities or specific natural or cultural resources allow for adverse project impacts if those impacts are sufficiently identified and considered in the decision-making process. Some requirements may specify that a project implement certain mitigation measures or be implemented in accordance with an approval or permit from an agency responsible for protecting that resource. An outright prohibition on the use of a particular resource, except for *de minimis* impacts or under extraordinary conditions, is not common to other environmental requirements.

Requirements Applicable to "Federal Actions"

In the 1960s Congress debated legislative options to address potential adverse impacts associated with federal actions. An action may be deemed "federal" based on the role that a federal agency plays in a project's approval or funding. A project funded under the Federal-aid Highways program would generally be considered a federal action. Two laws applicable specifically to federal actions that significantly affect the environmental review process for highway project development are NEPA and the National Historic Preservation Act (16 U.S.C. §470, et seq.).

As discussed previously, NEPA has two primary aims—to require federal agencies to consider the environmental impacts of a project and to give the public a meaningful opportunity to learn about and comment on the proposed project *before* a final decision is made. It is a procedural statute. That is, NEPA requires federal agencies to implement procedures to ensure that environmental impacts of a project are included among, but not elevated above, other factors considered during the federal decision-making process. If the adverse environmental impacts of the proposed action are adequately identified and evaluated, the agency is not constrained by NEPA from deciding that other benefits (e.g., community and economic benefits) outweigh the environmental costs and moving forward with the action. (The NEPA compliance process is discussed under "Demonstrating Compliance with NEPA.")

The National Historic Preservation Act (NHPA) declared a national policy of historic preservation to protect districts, sites, buildings, structures, and objects significant to American architecture, history, archaeology, and culture. NHPA did not mandate preservation of historic resources or prohibit adverse impacts to them, but Section 106 requires all federal agencies to consider the impacts of a proposal prior to taking any action that may affect a site included in, or eligible for inclusion on, the National Register of Historic Places.

NHPA also requires federal agencies to afford the Advisory Council on Historic Preservation (an independent federal agency created by the law) a reasonable opportunity to comment on federal actions that would affect properties on or eligible for inclusion on the National Register of Historic Places. For federally funded highway projects, FHWA must consult with the Advisory Council or the designated SHPO to determine project impacts to historic sites and potential ways to mitigate those impacts.

There are similarities between requirements established under Section 4(f) and Section 106, but also important differences between the statutes. Like NEPA, Section 106 establishes a procedural requirement that directs all federal agencies only to consider project impacts on certain resources. Section 4(f) applies only to DOT projects and prohibits the use of certain resources for those projects, except under certain conditions.

Additional federal laws and executive orders apply explicitly to federal actions that affect certain resources or communities. For example, a federally funded highway project may require compliance with additional requirements applicable to federal actions if that project may:

- involve the acquisition, rehabilitation, or demolition of real property that will displace persons from their homes, businesses, or farms as protected under the Uniform Relocation Assistance and Real Property Acquisition Act of 1970 (42 U.S.C. §4601, et seq., more commonly referred to as the Uniform Act);
- affect wetlands or floodplains pursuant to Executive Order 11990 or Executive Order 11988, respectively;
- convert farmland to nonagricultural uses pursuant to the Farmland Protection Policy Act of 1981 (7 U.S.C. §4201 et seq.);
- cause disproportionately high and adverse impacts on minority and low-income populations with respect to human health and the environment pursuant to Executive Order 12898; or

- affect human remains and cultural material of Native American and Hawaiian groups pursuant to the Native American Grave Protection and Repatriation Act (25 U.S.C. §3001 et seq.).

Requirements Applicable to Certain Resources

In addition to requirements applicable to federal-aid highways, specifically, and federal actions, in general, Congress has enacted a host of individual statutes intended to protect certain natural, environmental, and cultural resources from human-induced activities. A potentially long list of federal compliance requirements *could* apply to a given highway or bridge project, but requirements that will *actually* apply to a project will be limited by site-specific conditions and the degree to which the proposed project may affect protected resources. Broadly, highway projects may be subject to requirements intended to identify, minimize, or control adverse impacts to:

- **Land**—including land use that may affect the habitat of threatened or endangered plant and animal species, migratory birds, archaeological sites, and land designated as a national trail or national wilderness; and
- **Water resources or water quality**—including projects that may affect wetlands, aquatic ecosystems, navigable waters (e.g., rivers, streams, harbors), floodplains, coastal zones, or designated "wild and scenic" rivers, or projects that may affect water quality (e.g., discharge pollutants into U.S. waters).[17]

For a given federally funded highway project, compliance with a number of federal, state, or tribal regulations intended to identify, control, mitigate, or minimize project impacts to land and water resources may be required. Specific compliance requirements will depend on standards or regulatory requirements of that law and the degree to which the proposed project may adversely affect that resource. Depending on those factors, project development and implementation may require some level of consultation, analysis, or approval from an agency with jurisdiction over the resource. For example, a highway or bridge project that results in pollutants being discharged into wetlands, rivers, or streams or that may affect navigable waterways or harbors likely will require project development be completed in accordance with provisions established under the Clean Water Act or the

Rivers and Harbors Act. Pursuant to those laws, the selection of a particular project alternative may require a permit or certification from the Army Corps of Engineers (the Corps), the Environmental Protection Agency (EPA), the United States Coast Guard, or a state or tribal water quality control agency.

IMPLEMENTING THE ENVIRONMENTAL REVIEW PROCESS

The individual requirements discussed above were enacted by Congress after a particular concern arose or need was identified. For an individual project, several requirements involving similar compliance directives could apply. For example, depending on its impacts, a project may be subject to different public hearing or notification requirements under separate federal regulatory or statutory requirements.

The environmental review process is intended to function as the mechanism under which potentially duplicative requirements are identified and coordinated (including duplicative state or tribal requirements). Specifically, it is DOT policy that, to the fullest extent possible, any investigation, review, and consultation necessary to demonstrate environmental compliance be coordinated as a single process. The environmental review process is that single process. It forms the framework under which *all* applicable compliance requirements intended to protect the human, natural, or cultural environment are identified and demonstrated. Further, the NEPA compliance process forms the framework for completing the environmental review process.

In the past, suggestions made by transportation stakeholders to expedite project delivery, as well as legislative options proposed by some Members of Congress, have focused on requirements established specifically under NEPA. However, examples of individual projects delayed by environmental requirements more often involve issues associated with environmental compliance obligations established under separate state or federal requirements. In identifying and determining the potential effectiveness of nationally applicable approaches to expedite the environmental review process, it is necessary to distinguish between what is required explicitly under NEPA versus other federal environmental requirements.

Demonstrating Compliance with NEPA

The Council on Environmental Quality (CEQ) promulgated regulations implementing NEPA that were broadly applicable to all federal agencies.[18] CEQ required each federal agency to develop its own NEPA procedures specific to typical classes of actions undertaken by that agency.[19] In 1987, DOT promulgated "Environmental Impact and Related Procedures."[20] Those regulations prescribe the policies and procedures for FHWA to implement NEPA as it may apply to federally funded highway projects. They also include procedures necessary to ensure compliance with environmental requirements established under Title 23 applicable to Federal-aid Highways (e.g., procedures necessary to demonstrate compliance with requirements applicable to economic, social, and environmental effects, public hearings, and preservation of parklands (Section 4(f))). DOT's regulations have been revised periodically in accordance with legislative directives from Congress and to reflect court decisions applicable to DOT's implementation of both the NEPA process and its other environmental compliance obligations. Most recently, the regulations were modified to reflect the new environmental review process established under SAFETEA.

Identifying the Appropriate NEPA Document

For a given highway project that receives funding or approval under Federal-aid Highways programs, compliance with NEPA is demonstrated in the "NEPA document." Requirements that define the various categories of NEPA document and required elements of each are found in the NEPA regulations promulgated by both CEQ and DOT.

Transportation projects vary in type, size, complexity, and potential to affect the environment. To account for the variability of potential project impacts, NEPA regulations establish three basic "classes of action" that dictate how NEPA compliance will be documented and implemented. Determining the appropriate NEPA document and level of environmental review and analysis necessary for that document is dependent upon the answer to the following question: "Will the proposed action have any significant environmental impact?" Answers to that question, and the corresponding NEPA documents, are as follows:

- **Yes.** Those projects require an Environmental Impact Statement (EIS) followed by a final Record of Decision (ROD).

- **Maybe.** When the significance of a project's impacts is not clear, an Environmental Assessment (EA) must be prepared to determine whether an EIS is necessary or a Finding of No Significant Impact (FONSI) is appropriate.
- **No.** Those projects are categorically excluded from the requirement to prepare an EIS or EA; as such, those projects are generally referred to as Categorical Exclusions (CEs or CATEX).

Pursuant to NEPA's aims, an evaluation of environmental impacts is required prior to commitment of federal resources. To meet that requirement, preparation of the NEPA document may begin in the project planning stage, but must be completed within the preliminary design and environmental review stage of project development. Generally, subsequent stages of project development (final design activities, property acquisition, or project construction) cannot proceed until the necessary NEPA document is complete and approved by FHWA.

FHWA-Approved Projects by NEPA Class of Action

Major highway projects that require an EIS are the most studied and discussed when there is debate over the time it takes to complete the NEPA process. Further, past legislative efforts to expedite the NEPA process have focused primarily on the NEPA process as it applies to EIS preparation. However, FHWA data from 1998 to 2007 show that approximately 4% of federal-aid highway projects approved under programs administered by FHWA required an EIS. Projects processed as a CE or with an EA/FONSI accounted for approximately 96% (see **Figure 1**).

More recent FHWA data illustrate a similar proportion of major new projects and smaller maintenance/rehabilitation projects. In FY2009, of the approximately 55,043 miles of roadway projects receiving federal-aid highway funds, approximately 50,166 miles (91%) involved reconstruction projects with no added roadway capacity, restoration and rehabilitation activities, or road resurfacing (i.e., projects likely to be processed as CEs). Approximately 4,877 miles of road construction projects involved new construction, relocation, or reconstruction with added capacity (i.e., projects likely to require preparation of an EA or EIS).[21]

In addition to representing a small number of overall projects, few projects currently being developed require an EIS. As of November 18, 2011, 10 states had no active projects that involved EIS preparation, 12 states and the District of Columbia and Puerto Rico were preparing 1, and 18 states were preparing between 2 and 5 (illustrated in **Figure 2**). Further, a significant number of active EISs (68 of 175 or 39%) were being prepared in just five states—California, Texas, North Carolina, Florida, and New York.

While projects requiring an EIS represent a small proportion of total projects and a small number of active projects being developed in each state, they are more likely to be high-profile, complex projects that affect sizeable populations and take years, even decades, from planning to construction. They may cost millions, or even hundreds of millions, of dollars. For example, data from 1998 to 2007 regarding FHWA funding allocation show that while projects processed as CEs generally represent 90% of projects approved, those projects accounted for approximately 76% of FHWA program funds. Over that period, projects requiring an EIS accounted for approximately 4% of the total projects approved, but 12% of allocated program funds (see **Figure 3**).

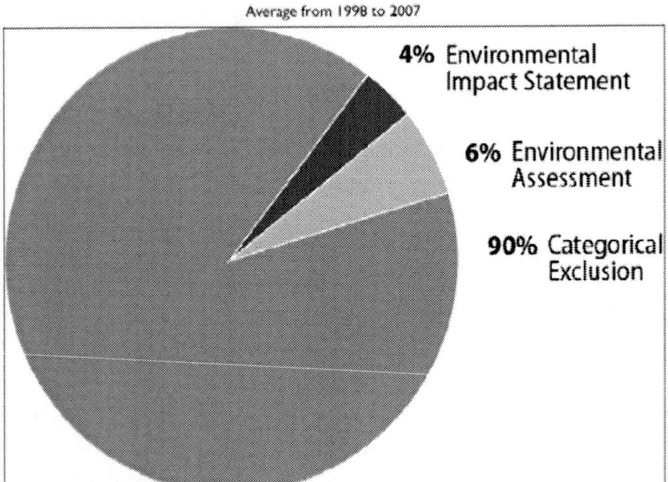

Source: Congressional Research Service, based on data available from FHWA's "Streamlining/Stewardship: Performance Reporting" website at http://www.environment.fhwa.dot.gov/strmlng/projectgraphs.asp/.

Figure 1. FHWA-Approved Projects—By NEPA Class of Action.

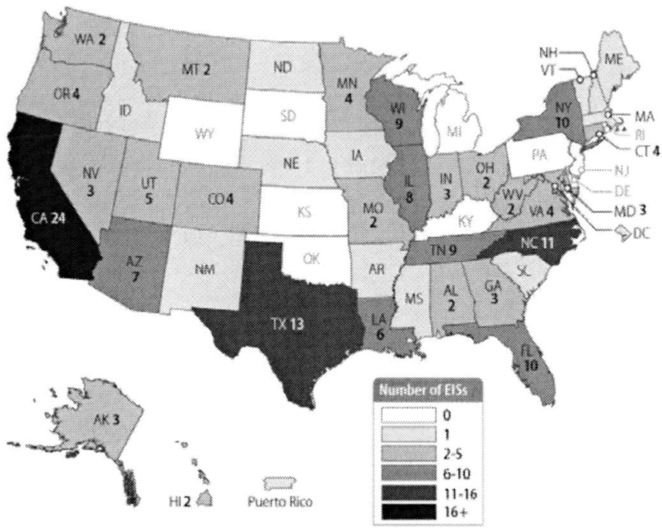

Source: Congressional Research Service, based on FHWA's list of Active and Inactive Environmental Impact Statements (EISs) as of November 18, 2011, available on the agency's "NEPA and Project Development" website at http://www.environment.fhwa.dot.gov/projdev/active_eis.asp.

Figure 2. Active FHWA Projects Requiring an EIS in Each State.

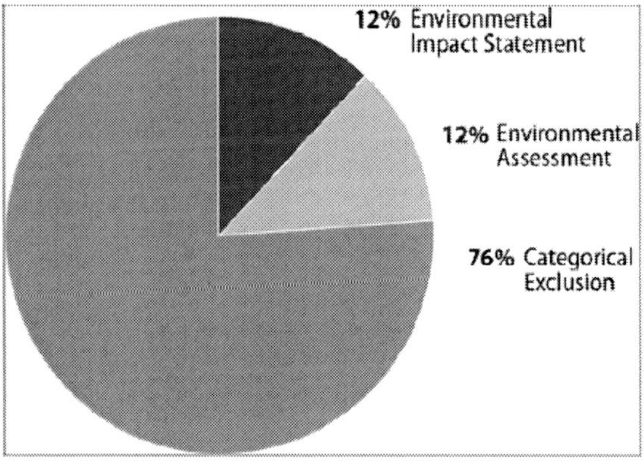

Source: Congressional Research Service, based on data available on FHWA's "Streamlining/Stewardship: Performance Reporting" website, http://www.environment.fhwa.dot.gov/strmlng/projectgraphs.asp/.

Figure 3. Percentage of FHWA Program Funding Allocation by NEPA Class of Action.

While a project requiring an EIS will likely cost more than a project processed as a CE, there is not necessarily a direct relationship between a project's cost and its level of environmental impacts. For example, it cannot be stated that projects that cost over $1 million, or even $10 million, will require an EIS. This point is illustrated by reviewing FHWA's list of "Major Projects," defined to include those expected to receive over $500 million in federal assistance.[22] Included on FHWA's list of currently active major projects are several that are being processed as CEs or with an approved FONSI.[23] For example, the "Loop 12/State Highway 35E Corridor" project in the Dallas-Fort Worth, TX, area is described as a reconstruction and widening project estimated to cost $1.6 billion. Project letting for that project began after approval of an EA/FONSI. Also included on the list is the "I-595 Corridor Improvements Project." That project, determined to be a CE, will add reversible lanes and involve major interchange improvements along 10.5 miles of the I-595 corridor in Florida. It is estimated to cost $1.8 billion.

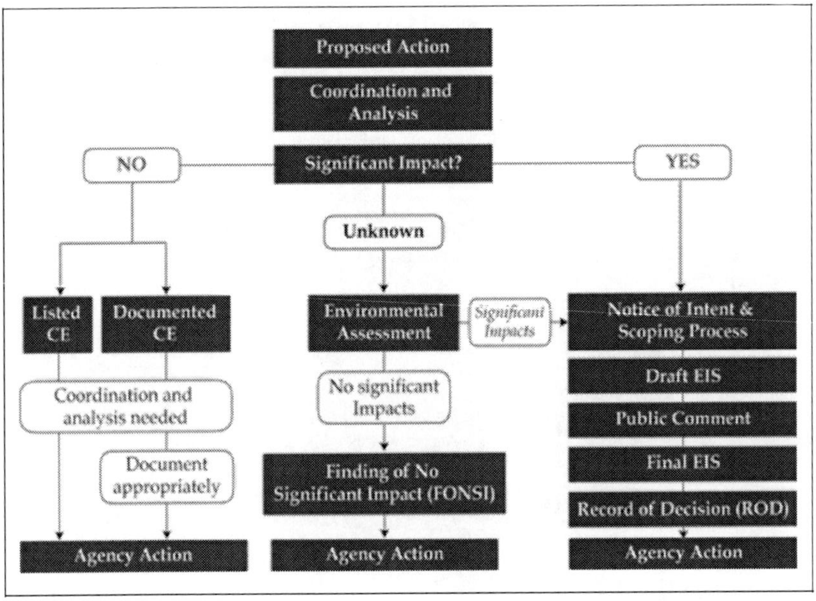

Source: FHWA guidance document, *Integrating Road Safety into NEPA Analysis: A Primer for Safety and Environmental Professionals*, in the "FHWA Environmental Toolkit," available at http://www.environment.fhwa.dot.gov/projdev/pd6rs_primer_sec2.asp.

Figure 4. NEPA Decision-Making Process.

Selected Requirements for Each Category of NEPA Document

Each NEPA document (EIS, EA, and CE) must include certain required elements (see **Figure 4**). That is, the NEPA document must show that environmental impacts were considered as part of the federal decision-making process, not a paperwork exercise to document impacts from a project after a decision was made.

Requirements applicable to each element of each NEPA document, and how DOT requires an applicant for federal funds to demonstrate compliance with each element, largely evolved in the 20 years after NEPA was enacted. Those requirements are reflected in both CEQ and DOT regulations implementing NEPA. The evolution of the NEPA compliance process was also influenced by the courts. For example, the courts played a prominent role in determining issues such as what constitutes "significant" impacts, who must prepare an EIS, at what point an EIS must be prepared, and how adverse comments from agencies should be handled. Changes to required elements of the NEPA process, applicable to projects funded under DOT programs, are also made by Congress.

Selected requirements applicable to each category of NEPA document, including requirements established under SAFETEA, are discussed below.

Categorical Exclusion (CE) Determinations

As discussed above, projects processed as CEs represent the greatest proportion of projects approved for federal-aid highway funds. DOT defines CEs as actions that, based on past experience with similar actions, do not individually or collectively have a significant impact on any natural, cultural, recreational, historic, or other resource, or involve significant air, noise, or water quality impacts; and that will *not*

- induce significant impacts to planned growth or land use for the area;
- require the relocation of significant numbers of people;
- have significant impacts on travel patterns; or
- otherwise, either individually or cumulatively, have any significant environmental impacts.[24]

A project may meet these criteria, but still involve "unusual circumstances" that would require FHWA to ensure that a CE designation is appropriate. Unusual circumstances applicable to federally funded highway projects include substantial project controversy on environmental grounds; a significant impact on properties protected under Section 4(f) or Section 106 of the National Historic Preservation Act (NHPA); or inconsistencies with any federal, state, or local requirements relating to the environmental aspects of the action.[25]

DOT identifies two groups of surface transportation projects that would likely meet the CE criteria (absent any unusual circumstances applicable to the project). The first group includes specific actions that meet criteria applicable to CEs.[26] DOT has determined that these projects (presented in **Table 1**) will likely result in insignificant environmental impacts because they either do not involve or directly lead to construction or involve minor construction.

A proposed action included in this list may or may not require an applicant for federal funds to submit supporting documentation to FHWA. Necessary paperwork could range from a simple checklist to substantial documentation. The extent of paperwork or supporting documentation is directly related to the extent of the impacts and necessary analysis of those impacts. For example, construction of a bicycle path or installation of traffic signals in a historic district may require some level of compliance with Section 106 of the NHPA or Section 4(f).

The second group of CEs includes actions that past DOT experience has shown to have substantial, but generally not "significant," effects.[27] For this group, DOT regulations include "examples" of actions commonly approved by FHWA that may meet the regulatory definition of a CE (presented in **Table 2**). Such projects require the project sponsor to provide FHWA with documentation to confirm that the project does not involve "unusual circumstances" resulting in significant environmental impacts. Unlike specifically "listed CEs" (**Table 1**), the potential universe of "documented CEs" is not limited to projects identified by DOT. Instead, FHWA may approve a CE designation for any action as long as documentation is provided that demonstrates the project meets the regulatory definition of a CE.

Although they are excluded from the requirement to prepare an EIS or EA, CEs are sometimes incorrectly identified as being exempt from NEPA or having *no* environmental impacts. No *significant* environmental impact *under* NEPA does not mean the project has *no other* regulated environmental impacts. For example, to demonstrate that a project meets both the CE criteria and will comply with other environmental requirements, state DOTs routinely gather information regarding a CE's potential to

- involve work that requires highway traffic or construction noise abatement;

- be located within certain limits of a sole source aquifer or alter stream flow;
- involve the acquisition of more than minor amounts of temporary or permanent right-of-way;
- require a Section 4(f) evaluation or "an opinion of adverse effect" under Section 106 of NHPA;
- involve commercial or residential displacement;
- involve work in wetlands that would require a permit from the Corps; or
- be constructed in a county that lists federal threatened and endangered species.

A project may involve any one or more of these (or other) activities that will have some effect on the human or natural environment, yet have environmental impacts that do not rise to the level of "significant" under NEPA. However, the threshold of significant impacts is primarily relevant to NEPA compliance. Other laws intended to protect or mitigate impacts to natural or cultural resources will have their own compliance thresholds applicable to *that* law. FHWA approval of a project processed as a CE may be delayed if the project sponsor does not realize that its proposed project may be subject to compliance requirements in addition to NEPA.

Within its responsibilities to oversee the Federal-aid Highway program, FHWA typically establishes procedures with each state DOT regarding CE review and approval. In a given state, "listed CEs" generally require minimal documentation before FHWA approval. NEPA review for those projects would be included as part of FHWA's project oversight and approval obligations established under Title 23. For "documented CEs," FHWA either reviews the NEPA documentation as part of the project development process and any agreed-upon procedures as part of the project review and approval, or the state DOT does this review in accordance with a formal programmatic CE agreement established between FHWA and the state DOT.

A programmatic CE agreement sets forth specific project circumstances for which a CE could be processed, and maintains FHWA oversight and responsibility for the NEPA determination. A programmatic approach involves establishing a streamlined process for handling routine environmental requirements, commonly applicable to specific types of project (e.g., bridge maintenance or road resurfacing activities). It allows for repetitive actions to be considered on a programmatic basis rather than project by project. Established on a local, regional, or statewide basis, a programmatic CE may establish procedures for consultation, review, and compliance with one or more federal laws. FHWA suggests that, to the extent possible, state DOTs take a programmatic approach to CE determinations.

Table 1. Federally Funded Highway Projects Specifically Listed as CEs

Non-construction activities (e.g., planning, technical studies, or research activities).	Emergency repairs after a natural disaster or catastrophic failure.
Installing fencing, signs, pavement markings, small passenger shelters, and traffic signals that involve no substantial land acquisition or traffic disruption.	Deploying electronic, photonic, communication, or information processing systems to improve system efficiency or safety.
Altering a facility to make it accessible to elderly and handicapped persons.	Landscaping activities.
Implementing ridesharing programs.	Improving existing rest areas or truck weigh stations.
Scenic easement acquisition.	Installing noise barriers.
Activities in a state highway safety plan.	Constructing bicycle or pedestrian lanes or facilities.

Source: Congressional Research Service, taken from actions listed at 23 C.F.R. §771.117(c).

Table 2. Examples of FHWA-Approved Projects That May be Classified as a CE

Actions That May Be Designated a CE with Appropriate Documentation and FHWA Approval	
Highway modernization through resurfacing, restoration, rehabilitation, or reconstruction.	Bridge rehabilitation, reconstruction, or replacement.
Highway safety or traffic operations improvement projects.	New truck weigh station or rest area construction.
Approval for changes in access control.	Approval for disposal of excess right-of-way or for joint or limited use of right-of-way.
Acquisition of certain preexisting railroad right-of-way.	Land acquisition for hardship or protective purposes.
Construction of transportation corridor fringe parking facilities.	

Source: Congressional Research Service, taken from examples of actions listed at 23 C.F.R. §771.117(d).

Apart from its potential to enter into programmatic CE agreements with FHWA, state DOTs may assume FHWA responsibility for CE determinations. Pursuant to provisions in Section 6004 of SAFETEA,[28] FHWA may assign and a state DOT may assume responsibility for determining whether certain highway projects meet the CE criteria. Under that authority, a participating state would be authorized to determine all CE applicability, including determining whether proposed projects that are not specifically listed under DOT's NEPA regulations may meet the CE criteria.

States that choose to assume FHWA responsibility would be required to do so in accordance with terms and conditions established in a memorandum of understanding (MOU) between the state and FHWA.[29] States assuming federal authority are legally liable for the NEPA determination. That is, FHWA would not be liable for the NEPA determinations for CEs in states participating in the program. FHWA would be required to conduct an annual review of a participating state's process for making CE determinations. To date, three states (Alaska, California, and Utah) have requested and been assigned responsibilities under the Section 6004 program.

Environmental Impact Statements (EIS)

Projects requiring an EIS make up the smallest percentage of projects approved by FHWA, but have generated the most attention when debating NEPA's potential role in delaying highway projects. FHWA does not specifically identify actions that require an EIS. That determination must be made on a case-by-case basis. However, DOT identifies the following as examples of highway projects that normally require an EIS: a new controlled access freeway; a highway project of four or more lanes on a new location; and new construction or extension of a separate roadway for buses or high-occupancy vehicles not located within an existing highway facility.[30]

Both the steps to complete an EIS and the EIS itself include certain required elements. Each required element represents the evolution of NEPA compliance requirements—as established by CEQ and, in part, as a result of judicial interpretation of NEPA's mandate and how its procedural requirements must be implemented. Required components in EIS preparation are

- file a Notice of Intent (NOI)
- scope the environmental issues
- prepare a draft EIS
- circulate the draft EIS for comment

- prepare the final EIS
- issue a final record of decision (ROD)

The NOI serves as the formal announcement of the project to the public and to interested federal, state, tribal, and local agencies.[31] As soon as possible after, or in conjunction with, the determination that an EIS is needed, the agency is required to determine the scope of the project. During that process, the project sponsor/applicant for federal funds should determine which environmental laws, regulations, or other requirements may apply to the project. During the scoping process, routes that may pose certain challenges and could be avoided may be identified (e.g., the presence of terrain or resources that may involve potential engineering or technical problems, regulatory restrictions, or public opposition). For example, during the scoping process, a potential route or alignment may be identified that would avoid property of historical significance, endangered species habitat, or wetlands—each of which may require compliance with the NHPA, the Endangered Species Act, or the Clean Water Act, respectively.

Once the scope of the action and its environmental issues have been determined, EIS preparation can begin. Required elements of an EIS, including selected elements in DOT's NEPA regulations or FHWA policy, are summarized in **Table 3**.

The EIS is prepared in two stages, resulting in a draft and a final EIS.[32] Supplemental documents may be required in some instances. Among other requirements, the final EIS must identify the preferred project alternative; reflect an evaluation of all reasonable alternatives considered; identify and respond to public and agency comments on the draft EIS; and summarize public involvement. The final EIS should document compliance with requirements of all applicable environmental laws, executive orders, and other related requirements. If full compliance is not possible by the time the final EIS is prepared, it should reflect consultation with the appropriate agencies and provide reasonable assurance that the requirements will be met. FHWA approval of the environmental document constitutes adoption of any findings and determinations in the EIS. The ROD presents the basis for the agency's final decision and summarizes any mitigation measures that will be incorporated in the project.

Each required element of the EIS involves compliance requirements established under both NEPA and other environmental requirements. For example, a clear delineation of project purpose and need is also necessary to meet the requirements under Section 4(f), executive orders on wetlands and floodplains, and permitting requirements established under Section 404 of the Clean Water Act. Identifying the potentially affected environment and analysis of environmental consequences also demonstrate that environmental impacts are considered during, not after, the decision-making process (as required under NEPA), but also include consultation, analysis, or input from resource agencies that may be necessary to ensure compliance with other applicable environmental law.

SAFETEA included several provisions that applied to projects that require an EIS. Section 6002 amended Title 23 by adding "Efficient Environmental Reviews for Project Decision-making" (§138). It established a new environmental review process applicable to all highways, transit, and multi-modal projects requiring an EIS. Among other requirements, the new process

- requires the project sponsor to notify DOT of the type of work, termini, length, general location of the proposed project, and a statement of any anticipated federal approvals;
- establishes a new entity required to participate in the NEPA process, referred to as a "participating agency," which includes any federal, state, tribal, regional, and local government agencies that *may have an interest in* the project;[33]
- requires the lead agency to establish a plan for coordinating public and agency participation in and comment on the environmental review process for a project or category of projects;
- requires the lead agency to establish a 60-day deadline on agency and public comments on a draft EIS and a 30-day deadline on all other comment periods in the environmental review process, except under certain circumstances (e.g., the deadline is extended by the lead agency for "good cause"); and
- prohibits claims seeking judicial review of a permit, license, or approval issued by a federal agency for highway or transit projects unless they are filed within 180 days after publication of a notice in the *Federal Register* announcing the final agency action, unless a shorter time is specified in the federal law under which the judicial review is allowed (previously, the six-year limit under the Administrative Procedure Act applied to NEPA-related claims).

DOT has produced guidance to help state DOTs implement SAFETEA's revised environmental review process and modified regulations implementing NEPA to reflect SAFETEA's amendments to Title 23.[34]

Table 3. Required Elements of an EIS as Implemented by FHWA

Elements of an EIS	Definition/Description
Purpose and need statement (§1502.13)	A brief statement, developed by the lead agency, specifying the underlying purpose of a project and the need to which the agency is responding. According to FHWA, this section may be the most important, as it establishes why the agency is proposing to spend large amounts of taxpayers' money while at the same time causing significant environmental impacts. A clear, well-justified statement explains to the public and decision makers that the use of funds is necessary and worthwhile, particularly as compared to other needed highway projects. The statement forms the basis on which potential alternatives to meet that need are identified and a final alternative is ultimately selected. It cannot be so narrow that it effectively defines competing "reasonable alternatives" out of consideration. The "purpose" may be a discussion of the goals and objective. The "need" may be a discussion of existing conditions that call for some improvement, including those applicable to transportation demand, safety, legislative direction, urban transportation plan consistency, modal interrelationships, system linkage, and the condition of an existing facility.
Alternatives (§1502.14)	Defined as the "heart" of the EIS, this section includes the identification and evaluation of all reasonable alternatives that may meet a project's purpose and need. FHWA requires the range of alternatives to include a discussion of how and why all reasonable alternatives were selected for consideration, and to explain why other alternatives were eliminated from detailed study. Each alternative, and its associated impacts, must be evaluated in sufficient detail to allow decision makers and the public an opportunity to compare the merits of each option.
Affected environment (§1502.15)	A succinct description of the environment of the area(s) to be affected or created by the alternatives under consideration. DOT requires this section to include a description of the existing social, economic, and environmental setting of the area potentially affected by all alternatives presented in the EIS. Data may include demographics of the general population served by the proposed project, as well as an identification of socially, economically, and environmentally sensitive locations or features in the proposed project area.

Elements of an EIS	Definition/Description
	For example, the EIS should identify the presence of affected minority or ethnic groups, parks, hazardous material sites, historic sites, or wetlands, among other factors.
Environmental Consequences (§1502.16)	Analysis of impacts of each project alternative on the affected environment, including a discussion of the probable beneficial and adverse social, economic, and environmental effects of each alternative. Where applicable, this section must include a description of the measures proposed to mitigate adverse impacts and methods of compliance with applicable legal requirements. FHWA recommends this section be devoted largely to a scientific analysis of the direct and indirect environmental effects of the proposed action relative to each alternative. Potential environmental consequences identified by FHWA include land use, farmland, social, economic, air quality, noise, water quality, wetland, wildlife, floodplain, or construction impacts; the requirement to obtain any permits; impacts to wild and scenic rivers, coastal barriers, threatened or endangered species, historic and archeological preservation, or hazardous waste sites; and any irreversible and irretrievable commitment of resources. This section would likely require input from other federal, state, tribal, or local agencies with expertise on the environmental consequences under review.
List of preparers (§1502.17)	List of names and qualifications of individuals responsible for preparing the EIS. FHWA requires this section to include lists of state and local agency personnel, including consultants, who were primarily responsible for preparing the EIS/performing environmental studies and FHWA personnel responsible for EIS preparation/review.
Appendix (§1502.18)	Prepared if necessary. An appendix normally consists of material that substantiates analysis fundamental to the impact statement.

Source: Congressional Research Service, taken from CEQ regulatory definitions under 40 C.F.R. §1502 and requirements and definitions applicable to highway projects included in FHWA guidance "NEPA and Transportation Decisionmaking: The Importance of Purpose and Need in Environmental Documents," and "Guidance for Preparing and Processing Environmental and Section 4(f) Documents."

The National Cooperative Highway Research Program conducted a survey of state DOTs to determine their impressions of the new environmental review process established by SAFETEA. The DOTs responding to the survey were generally favorable regarding the act's requirements.[35] In particular, there was wide approval of the 180-day statute of limitations.[36] However, survey respondents expressed concerns about some provisions, including their impressions that it represented no major change from what state DOTs were doing previously; it duplicated existing coordination procedures; and DOT already involved outside agencies prior to implementing the new procedures. Further, many survey respondents expressed concern that some requirements of the new environmental review process seemed to run counter to streamlining initiatives by creating additional requirements that could have a negative impact on schedules and budgets.[37]

Under Section 6005, SAFETEA amended Title 23 to establish a "Surface Transportation Project Delivery Pilot Program" (§327). The pilot program allowed Oklahoma, California, Texas, Ohio, and Alaska to assume certain federal environmental review responsibilities (in addition to the assumption of CE determinations established under Section 6004, discussed above). Responsibility could be assumed for environmental reviews required under NEPA, or any federal law, for one or more highway projects within the state. As a condition of assuming federal authority, Congress required the state to waive its right to sovereign immunity against actions brought by citizens in federal court and consent to the jurisdiction of federal courts. That is, the state would become solely liable for complying with and carrying out the federal authority that it consents to assume.

To date, only California has agreed to and developed a program to participate in the pilot program. Other states declined, primarily due to state legislature concerns regarding the potential liability associated with assuming federal responsibility for NEPA.

Additionally, some state transportation agency officials and stakeholders with an interest in transportation project development have expressed concern over DOT requirements implementing the pilot program (as required pursuant to the directive in Section 6005). Those objections have centered largely around DOT's requirement applicable to rights-of-way (ROW) acquisitions in states that choose to assume federal authority under NEPA. As discussed earlier, one of NEPA's primary aims is to ensure that federal agencies consider the impacts of their actions before proceeding with them. The NEPA process cannot simply document a decision that has already been made. This requirement means that federal funds cannot be used for ROW acquisitions (an action that could indicate that a final project decision has been made) before the NEPA process is complete. Currently, states may make ROW acquisitions using state funds on an at-risk basis. That is, they may purchase land using state funds, but risk losing future federal funding for that purchase if the project ultimately involves an alternative that does not use that property. By assuming DOT's authority, a state would assume federal agency-level responsibility to comply with NEPA. DOT has found that would mean, in its capacity as a federal agency, the state would be precluded from making such advanced ROW acquisitions.

Some have argued that the loss of a state's ability to make at-risk ROW acquisitions has been a disincentive to states that may otherwise want to assume federal authority under NEPA. However, when the fear of taking on federal liability and subjecting the state to the jurisdiction of federal court were primary reasons that states did not want to assume NEPA authority, it is unclear how states could be protected from potential judicial review if they are allowed to complete a transaction that *could appear* to violate one of NEPA's primary goals. Although state DOTs may be willing to accept that risk, a state legislature may not, particularly when an incentive for a state to assume the federal role under NEPA is to eliminate FHWA's oversight of the NEPA process (e.g., FHWA's legal sufficiency review of an EIS).

Environmental Assessments (EAs)

The third category of NEPA document is an EA. It is required for an action that is not a CE and does not clearly require an EIS, or where FHWA believes an EA would assist in determining the need for an EIS. An EA is intended to be a concise public document that serves to provide sufficient evidence and analysis for determining whether to prepare an EIS or a Finding of No Significant Impact (FONSI).[38]

In preparing an EA, the applicant, in consultation with FHWA, is required to consult with interested agencies at the earliest appropriate time to determine the project scope; determine which aspects of the proposed action have potential for social, economic, or environmental impact; identify alternatives and measures which might mitigate adverse environmental impacts; and identify other environmental review and consultation requirements which should be performed concurrently with the EA.[39]

The EA is subject to FHWA approval before it is made available to the public. The document itself need not be circulated, but must be made available for public inspection and comment (typically for at least 30 days). A notice of availability must be sent to state- and area-wide clearinghouses and should be published locally. Depending on FHWA-approved state procedures, a public hearing may or may not be required.

FHWA requires the basis of a request for a FONSI be clearly and adequately documented. Like an EIS, the EA or FONSI is required to clearly document compliance with NEPA and all other applicable environmental laws, executive orders, and related requirements. An approved FONSI functions as the final agency decision on a project.

Like projects processed as CEs, determining the time it takes to complete an EA is difficult. Local and state transportation officials do not routinely, nor could they easily, track the time it takes to complete an EA. A distinct end point could be identified (issuance of a FONSI), but a starting point may be hard to identify. Further, since EAs likely require limited environmental review or analysis under NEPA, any analysis or review that is prepared to support a FONSI would likely be required under separate state or federal law. However, transportation agency officials have complained that EAs sometimes approach the length of an EIS. If that is the case, factors indirectly related to the NEPA compliance likely apply to the project. For example, a project that may involve local controversy or opposition, but still have no significant impacts, may require more analysis or documentation than anticipated. Also, a project with substantial environmental impacts to certain resources may require time-consuming consultation, analysis, or approvals from agencies outside DOT to confirm that no significant impacts will occur, or it could be an indication that an EIS should have been prepared initially.

Agency Roles and Responsibilities in the NEPA Process

The NEPA document is prepared by a "lead agency," and may require input and analysis from "cooperating" or "participating" agencies. Depending on the environmental impacts of a given project, both the lead and cooperating agencies are obligated to meet certain federal requirements. The time it takes to meet those obligations has been identified by transportation agencies as a potential source of frustration or project delay.

Lead Agencies

The "lead agency" is the federal agency responsible for preparing the NEPA document.[40] DOT must serve as the lead federal agency for a federally funded transportation project (FHWA generally serves as the lead for highway projects). The direct recipient of federal funds for the project must serve as a joint lead agency (a requirement explicitly established under SAFETEA). For a federal-aid highway project, that is typically the state DOT, but may include a local agency project sponsor or a federally recognized Indian tribal governmental unit. At the discretion of the required lead agencies, other federal, state, or local governmental entities may act as joint lead agencies. These include, but are not limited to toll, port, and turnpike authorities and metropolitan planning organizations (MPOs). For example, the U.S. Department of Homeland Security may serve as a joint lead agency with DOT and the project sponsor on a transportation improvement at a national border crossing.

In practice, the entity seeking federal funds will prepare the NEPA document, and other supporting environmental review documents, with guidance from FHWA (as necessary or as requested). FHWA, however, has ultimate responsibility to ensure that a project seeking federal funds will comply with the various laws, regulations, and executive orders applicable to the project. In that capacity, before final approval and project funding, FHWA is required to independently evaluate the necessary environmental documents and review the legal sufficiency of a final EIS[41] or Section 4(f) evaluation.[42] This review is intended to ensure that the Section 4(f) evaluation or NEPA document is consistent with legal requirements. It includes a review of the documentation and associated compliance efforts to determine if those efforts are sufficient to assure compliance with applicable law. A separate technical review of the final NEPA/Section4(f) document is also conducted by FHWA, prior to document approval.

Cooperating and, after SAFETEA, Participating Agencies

The lead agency must consult with and obtain the comments of any federal agency that has "jurisdiction by law or special expertise with respect to any environmental impact involved" in an action that requires an EIS.[43] In CEQ's NEPA regulations, those agencies are identified as "cooperating" agencies.[44] Pursuant to directive from Congress in SAFETEA, DOT's NEPA regulations were supplemented to also identify "participating" agencies, which may include any federal and non-federal agencies that may have an interest in the project.[45]

At the request of the lead agency, the cooperating agency is required to assume responsibility for developing information and preparing environmental analyses, including portions of the EIS related to its special expertise. Cooperating agencies are also obligated to provide comments on the NEPA document on areas within their jurisdiction, expertise, or authority. For projects requiring an EIS, that role may be set out in a memorandum of understanding or agreement between the agencies. The lead agency is also required to request comments from appropriate state, local, or tribal agencies;

any agency that has requested to receive EISs on similar actions; and the project applicant.[46]

CEQ regulations specify requirements for inviting and responding to comments on the draft EIS (including requirements that specify a cooperating agency's duty to comment on the draft).[47] The lead agency is required to consider those comments and respond in one of the following ways:

- modify proposed alternatives, including the proposed action;
- develop and evaluate alternatives not previously considered;
- supplement, improve, or modify its analyses;
- make factual corrections in the EIS; or
- explain why the comments do not warrant further response from the lead agency, citing the sources, authorities, or reasons that support its position.[48]

As illustrated in the choices listed above, the lead agency is not precluded from moving forward with a project if it explains why a cooperating agency's comments do not warrant further response. However, FHWA suggests that every reasonable effort be made to resolve interagency disagreements on actions before processing the final EIS. If significant issues remain unresolved, the final EIS shall identify those issues and any consultation or other effort made to resolve them.

Some highway projects have involved disagreements regarding the appropriate authority and extent of involvement of coordinating agencies in the NEPA process. For example, in 2003, Transportation Secretary Norman Mineta requested CEQ Chairman James Connaughton to clarify the role of lead and cooperating agencies in developing EIS statements of project purpose and need statements.[49] Secretary Mineta cited the sometimes lengthy interagency debates over those statements as a cause of delay in highway project development. In his response, Chairman Connaughton referred to CEQ regulations specifying that the lead agency has the authority and responsibility to define a project's purpose and need. Further, Chairman Connaughton referenced previous federal court decisions giving deference to the lead agency in determining a project's purpose and need. Chairman Connaughton's letter also quotes CEQ's regulations, citing the lead agency's "responsibilities throughout the NEPA process for the 'scope, objectivity, and content of the entire statement or of any other responsibility' under NEPA."

Public Involvement

To meet NEPA's goal applicable to public participation in federal decision making, CEQ's regulations require agencies to provide public notice of NEPA-related hearings, public meetings, and the availability of environmental documents so as to inform public stakeholders.[50] DOT procedures extend beyond those established under CEQ regulations to reflect requirements applicable to "public hearings" established under Title 23.[51] For example, EAs do not need to be circulated, but must be made available to the public through notices of availability in local, state, or regional clearinghouses, newspapers, and other means. Depending on a state's public involvement procedures (approved by FHWA), a public hearing may or may not be required for projects that proceed with an EA. Pursuant to DOT regulations implementing NEPA, documentation necessary to demonstrate compliance with Title 23's public hearing requirements (e.g., public comments or hearing transcripts) must be included in the final EIS or FONSI, as applicable.

Stakeholders that comment on surface transportation projects may be expected to vary depending on a project's impacts. They may include individuals or groups who may benefit from or be adversely impacted by the project, or special interest groups with concerns about the project's impacts on certain affected environments. For example, a highway project that involves upgrading existing roadways may involve construction activities that would affect adjacent homes or businesses. The project may elicit comments from the local business community (e.g., individual businesses, the Chamber of Commerce, or local development organizations) or area homeowners. A project that may affect sensitive environmental resources, such as wetlands or endangered species, may generate comments from local or national environmental organizations.

If a member of the public has concerns about a project's impacts, comments may be directed at virtually any element of the NEPA process or related documentation. Someone may disagree with the definition of project's purpose and need discussion, the range of "reasonable" alternatives selected for consideration and analysis, or the identified level of significance of the project's impacts (e.g., a FONSI was issued when the individual felt an EIS should have been required). Issues that arise during the public comment period may also be the subject of legal action. Critics of NEPA charge that those who disapprove of a federal project will use NEPA as the basis for litigation to delay or halt that project. Others argue that litigation results only when agencies do not comply with NEPA.[52]

Actual litigation played a prominent role in NEPA's early implementation. However, it may be the *threat* of litigation that affects its current implementation. The number of NEPA-related lawsuits filed annually against FHWA is low.[53] Still, the potential threat of litigation may result in an effort to prepare a "litigation-proof" NEPA document. This may be the case particularly for projects that are costly, technically complex (potentially requiring compliance with multiple environmental laws), or controversial (e.g., opposed by or individuals affected by the project or groups that anticipate adverse impacts to resources of concern to them). Some look at this positively, asserting that the fear of a lawsuit makes agencies more likely to adhere to NEPA's requirements. Others counter that the threat of litigation may lead to the generation of wasteful documentation and analyses that do not add value to, and slow, decision making.

Demonstrating Compliance with Additional Requirements

Unlike NEPA, which will apply in some way to all federally funded highway projects, additional environmental requirements applicable to a project will depend on site-specific conditions and potential impacts to resources at the site. For example, what and how requirements may apply to a project will depend on its effect on water quality, water resources, and land use as well as community, visual, noise, or social impacts, to name a few. While a wide array of requirements may apply to federally funded highway projects, certain federal requirements apply more commonly than other requirements. Also, certain compliance requirements have been identified by transportation stakeholders as those more likely to delay the environmental review process (see surveys and studies listed in **Appendix**). The most commonly applicable laws, and selected compliance requirements, are listed in **Table 4**.

As illustrated by the requirements listed in **Table 4**, when a federal highway project involves regulated impacts to certain resources, an agency with jurisdiction over that resource may be required to provide some level of analysis, consultation, or approval before a project can proceed. Resulting consultation or approval may include directive(s) to the project sponsor regarding how or whether the proposed project may use the resource. These requirements can lengthen the time it takes to complete the overall environmental review process, if outside agency opinions, input, and/or evaluations are required before the NEPA review can be completed. Whether such requirements will lead to project delays could depend on a host of factors such as whether the project sponsor anticipated the need for outside agency approval or the workload of the agency processing the approval.

Table 4. Federal Law Commonly Applicable to FHWA-Approved Projects

Federal Law	Selected Compliance Requirements
Section 4(f)	For projects that would use a 4(f) resource, an evaluation or a determination of *de minimis* impacts must be prepared (see the discussion regarding "Preservation of parklands" in the "Requirements Applicable to Federal-aid Highways" section). The evaluation or *de minimis* impacts determination requires some level of consultation with or concurrence from the official with jurisdiction over the resource (e.g., the Department of the Interior's U.S. Fish and Wildlife Service (FWS); federal, state or local park authorities; or the designated SHPO).
Section 106 of the National Historic Preservation Act	For projects that may affect a site included, or eligible for inclusion, in the National Register of Historic Places, FHWA must consult with the Advisory Council on Historic Preservation (ACHP) or the designated SHPO to determine impacts to the site and seek ways to avoid, minimize, or mitigate adverse impacts. Affected parties must be involved in mitigation plans.
Endangered Species Act (Section 7)	FHWA must prepare a biological assessment when the presence of threatened or endangered animals or plants is suspected to occur in the vicinity of a project. FHWA must consult with the federal agency of jurisdiction (FWS or the Department of Commerce's National Marine Fisheries Service (NMFS)) that will issue a biological opinion on whether the proposed action would jeopardize the continued existence of listed species, or destroy or adversely modify their designated critical habitats.
The Clean Water Act (Section 404)	Requires that the discharge of dredge and fill materials into navigable waters of the United States be done in accordance with review and permitting procedures administered by the Corps, under guidelines developed by EPA. Other federal agencies potentially involved in permit evaluation process include FWS or NMFS.

Source: Congressional Research Service, taken from requirements listed in FHWA's "Summary of Environmental Legislation Affecting Transportation," at http:// www.fhwa.dot.gov/environment/env_sum.htm.

Note: This is not intended to be an exhaustive list of federal requirements potentially applicable to federally funded highway projects or a complete description of potentially applicable compliance requirements established pursuant to each law. However, the selected requirements illustrate the potential need to obtain permits or consult with agencies outside DOT. Further, federal laws selected for listing in this table represent those identified by transportation agency officials as a common source of delay in completing the environmental review process.

To integrate the NEPA compliance process and avoid duplication of effort associated with a project's overall environmental compliance obligations, CEQ's NEPA regulations specify that, to the fullest extent possible, agencies must prepare the NEPA documentation concurrently with any other environmental requirements. The appropriate NEPA documentation should demonstrate compliance with all applicable environmental requirements. It must indicate any federal permits, licenses, and other approvals required to implement the proposed project. This means that compliance requirements of any additional environmental laws, regulations, or executive orders must be identified (but not necessarily completed) during the NEPA process. If full compliance is not possible by the time the final NEPA document is prepared, the document should reflect consultation with the appropriate agencies and provide reasonable assurance that the requirements will be met.

Environmental Reviews and Project Development

To understand how the environmental review process may affect project delivery, it is useful to understand how the process fits into overall project development, as well as the challenges associated with measuring each stage of that development. It is also useful to recognize root causes of delay in completing the environmental review process, as well as how the process can lead to more efficient project development.

Stages of Project Development

Federal-aid highway funds are generally apportioned to each state by FHWA for the construction, reconstruction, and improvement of highways and bridges on eligible highway routes, and for other special-purpose programs. Individual state DOTs are responsible for determining how and on which projects those funds will be spent. In making that determination, multiple activities and decisions occur from the time a tribal or state DOT, metropolitan planning organization, or local program agency (such as a municipal public works agency) identifies a transportation-related need and a project addressing that need is constructed.

Each stage of project development is initiated and completed largely at the local, tribal, or state level, with ultimate project approval at the federal level—from FHWA for federally funded highway projects. Although the names and details of each step may vary from state to state, they generally include project planning, preliminary design and environmental review, final design and right-of-way acquisition, and project construction. Activities common to each phase of the project development process, including maintenance activities that may take place after project construction, are described in **Table 5**. The table also identifies potential environmental compliance obligations that may occur in each stage of project development.

Table 5. Stages of Federal-Aid Highway Project Development

Common Project Activities and Environmental Compliance Obligations in Each Phase of Development		
Project Phase	**Description/Common Activities**	**Common Environmental Compliance Obligations**
Planning	Transportation program or project planning involves a cooperative process designed to foster involvement by all users of the planned system—such as the business community, community groups, environmental organizations, the traveling public, freight operators, and the general public. During this stage, a proactive public participation process is conducted by the metropolitan planning organization (MPO), state DOT, and transit operators. Among other activities, MPOs and state DOTs identify current and projected future transportation problems and needs, and analyze, through detailed planning studies, various transportation improvement strategies to address those needs. They also develop long-range plans and short-range programs for alternative capital improvement and operational strategies for moving people and goods.	Efforts have been made, in both FHWA guidance and statutory directive from Congress, to link statewide and metropolitan planning to the environmental review process. For example, Section 6001of SAFETEA requires the development of long-range transportation plans to include consultations with resource agencies responsible for land-use management, natural resources, environmental protection, conservation, and historic preservation, which may involve comparisons of resource maps and inventories; discussion of potential mitigation activities; and participation plans that identify a process for stakeholder involvement.
Preliminary design and environmental review	A project applicant identifies the preliminary engineering issues, such as proposed alignment of roadways, costs, and project details. This stage includes preliminary engineering and other activities and analyses, such as topographic or metes and bounds	An applicant for federal-aid funds must determine the appropriate NEPA document to be prepared and identify various resources potentially affected by a proposed project and its alternatives. The final NEPA

Table 5. (Continued)

Project Phase	Description/Common Activities	Common Environmental Compliance Obligations
	surveys, geotechnical investigations, hydrologic or hydraulic analysis, utility engineering, traffic studies, financial plans, revenue estimates, hazardous materials assessments, and general estimates of the types and quantities of materials and other work needed to establish parameters for the final design.	document must identify and demonstrate compliance with any other applicable environmental requirement, to the maximum extent possible, including completing necessary environmental or engineering studies, outside agency coordination or approvals, and public involvement.
Final design and right-of-way acquisition	Final construction plans and detailed construction specifications for the selected project alternative are prepared. If necessary, property appraisals and the acquisition of rights-of-way (ROW) or property to mitigate environmental impacts are made. Property acquisition that may involve the relocation of residents and businesses must be done in accordance with the Uniform Act of 1970. Also, if necessary, utilities are relocated. Project costs are finalized.	Property or material purchases cannot proceed until the NEPA document is approved by FHWA. Property acquisitions must be completed in compliance with requirements identified in the document. If late changes to the project are required, the environmental review process may have to be revisited if design changes result in unanticipated or previously unidentified environmental impacts.
Construction	The state DOT, or other project sponsor, requests and evaluates bids, and awards contracts. Project construction must reflect decisions made during the planning, environmental review, and design stages of project development.	Necessary permits or other compliance requirements identified during environmental review must be in place. Mitigation measures must be complete (e.g., installation of noise barriers or implementation of wetland mitigation). If elements of the project change, the environmental review process may have to be revisited if changes result in unanticipated environmental impacts.
Maintenance	Although not considered part of project development, the majority of projects funded under FHWA-approved programs involve activities that may be broadly described as "maintenance." Highway maintenance may include modernization through roadway resurfacing, restoration, rehabilitation, reconstruction, or adding shoulders or auxiliary lanes. Bridge maintenance may include rehabilitation, reconstruction, or replacement.	Identifying, planning, and implementing necessary maintenance activities are likely initiated and carried out at the local level, with state DOT approval. Maintenance activities would commonly involve a CE determination as well as an assessment of impacts that may require compliance with additional environmental requirements (e.g., impacts to historic sites or structures or endangered species habitat).

Source: Congressional Research Service, based on a review of state DOT practices.

Frequently, "environmental review" is considered synonymous with "NEPA compliance." That is not the case. However, completion of the NEPA compliance process and the overall environmental review process are linked by DOT's requirement that a project cannot be approved and subsequent stages of project development cannot proceed until the project sponsor appropriately documents compliance with NEPA and other applicable environmental requirements.

Challenges in Measuring Stages of Project Development

There are distinct activities associated with each stage of project development. However, the following factors make it difficult to estimate the time it takes to complete each stage:

- Most state and local transportation agencies do not maintain a centralized source of data tracking the time it takes to complete transportation projects. Further, there is no acceptable measure of when a project is delivered in a timely manner versus delayed. A project or a stage of its development may be considered "delayed" if it took the project sponsor longer than anticipated.
- Most state and local transportation agencies do not attempt to extract and measure the time it takes to complete individual activities attributable to a single stage of development (e.g., activities categorized distinctly as applicable to "environmental review"). Further, tracking that data may be difficult since elements of one phase may overlap with another (e.g., project planning activities may include elements of environmental review) and a distinct start and end point of individual activities may be difficult to identify.
- Project development may start, stop, and restart for reasons unrelated to environmental compliance. For example, EIS preparation may begin with publication of a NOI, but preparation may stop and restart due to changes in state priorities, funding availability, or a host of other issues unrelated to NEPA. In such cases, the time between issuing a NOI and ROD are an inaccurate measure of the NEPA process.
- Differences between state DOT project development processes make it difficult to establish a nationally applicable measure of project development stages. Also, the influence of tribal- or state-specific

environmental compliance requirements makes it difficult to isolate the time it takes to comply with federal requirements.

Considering these points, it is difficult to determine either the time it takes to meet specific elements of the environmental review process (e.g., NEPA compliance or agency consultations under the Endangered Species Act) or the degree to which completing the process delays project delivery. Further, it is not possible to assert, with any degree of accuracy, broad, nationally applicable values to the time it takes to complete the environmental review process. For example, there are no data available to substantiate a statement such as "environmental compliance accounts for X% to Y% of surface transportation project development time," or "compliance with NEPA or Clean Water Act permitting requirements delays X% to Y% of projects for X to Y months/years." Instead, it may be possible to determine "bridge reconstruction or rehabilitation in state A takes from X to Y months/years" if state A is one that tracks such information.

Also, it may be generally stated that the time it takes to deliver larger, more complex or controversial projects takes longer to complete than is typical for the majority of FHWA-approved highway projects (e.g., maintenance and rehabilitation projects). In addition to taking longer to complete due to their potential cost, size, and complexity, they will likely require compliance with more state, tribal, and federal requirements and may generate more public interest or opposition.

In 2002, the General Accounting Office (GAO) released a report that attempted to determine the typical amount of time it takes to complete overall project delivery as well as individual phases of project development for certain federally funded highway projects.[54] Data for this report were compiled based on the professional judgment of FHWA staff, staff of state departments of transportation, and transportation associations. According to FHWA, planning, gaining approval for, and constructing federally funded major highway projects that involved new construction typically took from nine to 19 years from planning to construction. FHWA estimated that the preliminary design and environmental review phase for those projects typically took from one to five years, depending on the complexity of the design and possible environmental impacts that must be considered.[55] It was noted that projects studied in the GAO report included those that would typically require an EIS and represent a small percentage of federally funded projects. It was also noted that, while there are many reasons new highway construction projects may take a long time to complete, most studies on project delivery focused only on the timely resolution of environmental issues to improve project completion times, rather than examining all aspects of project development.[56]

Causes of Delay in Completing Environmental Reviews

Although the extent to which the environmental review process may delay project delivery is unclear, it is generally not disputed that the time it takes to demonstrate compliance with environmental requirements can be time-consuming, particularly in cases where EIS preparation is required. Also, while transportation agency officials may cite elements of the environmental review process as a source of frustration or delay, it is not clear what specific environmental compliance requirements currently and routinely lead to project delays or the root cause of those delays.[57] For example, a common complaint among transportation agency officials is that outside agencies (including FHWA review and approval of the final NEPA documents) do not provide necessary input or approval in a timely way. However, there is little information available that clearly indicates *why* that may be the case on anything other than a project-specific level.

Few studies have looked at the root causes of project delay directly attributable to the environmental review process. Available studies have looked at a limited number of major new construction projects that required an EIS. By their nature, those projects involve unique project-specific issues and are likely to involve complex design, engineering, and compliance issues. Causes of delay for those projects more likely represent the exception and not the rule.

A 2003 FHWA study that attempted to identify causes of delay in completing EISs was unable to identify common factors or conditions that directly or indirectly affected the time it took to complete the NEPA process.[58] Although timing varied by broad geographic region, it did not seem to vary in relation to the majority of other variables considered (e.g., the presence of certain "controversial issues" or the required participation of agencies outside DOT). Instead, it was observed that the time it took to complete the NEPA process may have been more affected by external social and economic factors associated with broad geographic regions of the country.[59]

Subsequent, albeit limited, study data and anecdotal evidence regarding individual projects also point to factors external to environmental reviews as those most likely to delay the process. In particular, causes of delay in completing environmental reviews arise primarily from potentially overlapping local and project-specific issues including, but not limited to, the following:

- **Local issues**—the project's level of priority among others proposed in the state; changes in funding availability; concerns of local property owners; or opposition to the project (which may or may not be connected to environmental issues).
- **Project-specific issues**—the project's technical complexity; changes in project scope or design; lawsuits or the threat of litigation (which may or may not be connected to environmental issues); poor consultant work; issues with city documentation; issues with new alignment or coordination with other transportation projects; or land use planning issues.[60]

As discussed previously, environmental requirements identified as a source of delay have been associated with selected requirements established under Section 4(f); the Endangered Species Act; the National Historic Preservation Act; and the Clean Water Act. If a project is delayed by requirements under those laws, that delay may be attributable to project-specific issues.

Efforts to identify specific problems or causes of delay in meeting requirements other than NEPA have found differing perceptions among resource agency and transportation agency officials. For example, in completing its obligations as part of the environmental review process, resource agencies have identified poor communication, problems with the project's alternative analysis, being given incorrect or incomplete information, disagreements or differences of opinion among agencies, or environmental or biological issues associated with the project. Transportation agency officials also cited disagreements or differences of opinion and environmental or biological issues associated with the project, but identified a lack of timely response from resource agencies as the primary problem.[61]

Benefits to the Environmental Review Process

When there is debate over potential options to expedite the environmental review process, that process may be viewed as simply an obstacle to overcome before a highway or bridge project can be built. Benefits to the process may be overlooked or hard to quantify. Potential benefits may generally be thought of as those associated with balancing transportation and infrastructure needs with environmental protection and community concerns. However, one benefit that is not often considered is the degree to which the environmental review process may ultimately save time and reduce overall project costs by identifying and avoiding problems in later stages of project development. A study prepared for the Transportation Research Board made this point when evaluating causes of delay in the construction phases of development.[62]

Among other findings, the study found that certain recognized management principles, identified as relevant to timely completion of highway construction projects, should be applied by state highway administrators and contractors. It found that adherence to these principles was often inconsistent and lacking, usually resulting in construction delays. Among the principles identified was the "Cost-Time Relationship," under which, the study found,

> More time spent in design identifying problems will reduce construction time and result in a shorter overall project time. A widely recognized principle is that spending more monies during planning and design will reduce the time and cost required for construction by avoiding unforeseen conditions, reducing to a minimum design errors and omissions, and developing schemes that will support the most efficient approach to construction. In the design phase, the opportunity to make decisions to influence the final project cost is greatest. Yet, the expenditure of project funds is comparatively minimal, typically about 10% of the capital budget.[63]

These study findings illustrate the potential problem with considering time spent in the planning or preliminary design stage as a delay. It is impossible to determine whether or how much time may be saved, and project delivery ultimately accelerated, by avoiding conditions identified early in the process.

EXPEDITING ENVIRONMENTAL REVIEWS

Lessons Learned

The potential for the environmental review process to expedite project delivery is illustrated in findings of a 2009 peer exchange between representatives from state DOTs and FHWA Division Offices. The exchange was intended to identify strategies to more effectively move complex environmental documents through the EIS process.[64] Participants presented information on projects in their states that had moved through the environmental review process quickly. They highlighted the challenges encountered, methods used to successfully and efficiently navigate the EIS process, and lessons learned from their experience. It was observed that the practices described by state DOTs represented a fundamental shift in the way agencies have conducted environmental reviews over the last 10 to 15 years. Those state DOTs were found to have

> embraced innovative and creative solutions to balance transportation and infrastructure needs with environmental protection and community concerns. The environmental review processes for the successful projects … were conducted in a collaborative and transparent manner, whereby [state DOTs] sought to include stakeholders early and often throughout development of the EIS. Such methods not only lead to a faster completion of the environmental review process, but perhaps more importantly, they result in the delivery of better quality projects, ones that fulfill the transportation needs of communities while maintaining protection of environmental resources at the same time.[65]

One recent event serves as a good example of how environmental compliance requirements can be coordinated efficiently. That event was the reconstruction of the I-35 bridge in Minneapolis after its August 1, 2007, collapse. A new bridge opened just over a year later on September 18, 2008. The timing of that bridge reconstruction led to the question "Why can't all projects be completed that quickly?"

The answer to this question can be found, in part, in an FHWA study that examined how the key elements of the environmental review process were completed after a bridge collapse.[66] A primary factor cited in the study was that, in the wake of an emergency, the major causes of surface transportation project delay are absent. The "major causes of delay" identified were a lack of funding or priority in the state for the project; local controversy; interested stakeholder or local opposition; or insufficient political support.[67] Other potential causes of delay could still apply to emergency projects, including issues with the projects' complexity, poor consultant work, or the environmental review process.

The FHWA study looked at the Minnesota bridge collapse as well as other projects that involved bridge reconstruction after a collapse. Projects in the study illustrated how efficiently the environmental review process *could be* implemented if the more common sources of delay are absent and environmental review involves efficient interagency cooperation.

Bridge reconstruction for the I-35 project required the same environmental permits that would apply to any bridge reconstruction project of similar scope and scale. Despite the urgency of the project, there was no waiver or exemption from the environmental review or permit requirements. The replacement bridge was widened to accommodate future transit options, but did not increase capacity. The project fit the criteria necessary to be processed as a CE, but still required

- a permit issued by the Corps under Section 404 of the Clean Water Act;
- a bridge construction permit issued by the U.S. Coast Guard;
- an assessment of potential impacts to threatened and endangered species by a consultation team formed by FHWA, Minnesota DOT (MnDOT), and the Department of the Interior's U.S. Fish and Wildlife Service (FWS);
- a Minnesota Pollution Control Agency permit certifying compliance with the Clean Water Act's National Pollutant Discharge Elimination System and other requirements; and
- an assessment of potential cultural and historic issues through MnDOT's Cultural Resources Unit (CRU), in part, in accordance with a programmatic agreement with the Minnesota SHPO and tribes interested in reviewing state projects.

Efficient interagency coordination on the project was a factor identified as one associated with expedited reconstruction of the bridge. However, the efficiency of that agency interaction did not begin with this project. FHWA observed that staff from state and federal agencies involved in the environmental review process had worked collaboratively on past projects. The agencies established lines of communication and understood the tasks and concerns of each other's agencies. Those existing relationships led to a quick response among those agencies after the bridge collapse. Further, FHWA and MnDOT recognized that by limiting the scope of the project, the environmental review process was expedited because no expanded environmental review was needed (e.g., it met the criteria applicable to a CE). Further, federal and state resources were focused on this project—its completion was a priority to the state.

Apart from issues cited in the FHWA bridge study, MnDOT cited its use of a design-build procurement process as an important factor in the expediting project completion. A "design-build" process brings designers and contractors together early in the project development process and allows for a shortened process completion time by overlapping design and construction.[68]

Lessons learned from projects completed relatively quickly as well as suggested solutions from transportation agency officials[69] involve certain common approaches or procedures that have or could streamline the environmental review process. Those approaches include the following:

- efficient interagency communication and project coordination;
- early and continued communication with stakeholders affected by a project;
- improvements in internal processes and procedures;
- demonstrated agency commitment to priority projects and project schedules; and/or
- programmatic approaches to meeting compliance obligations.

Each of these approaches can be implemented under existing standards and requirements applicable to federally funded highway projects. For example, CEQ and DOT regulations implementing NEPA include explicit requirements intended to identify potential environmental issues early in the project development process and coordinate efficient interagency cooperation. CEQ also provides federal agencies with guidance on improving the efficiency and timeliness of their environmental reviews under NEPA.[70] DOT provides guidance and information intended to assist state and local agencies in implementing the environmental review process more efficiently.[71]

Administrative Efforts

In 2009, DOT initiated its "Every Day Counts" program to identify and implement approaches to shortening project delivery (among other goals). The program includes an evaluation of potential changes in DOT's role in implementing the environmental review process, including the following efforts:

- **Improve the link between project planning and environmental review**—sets up a framework for incorporating planning documents and decisions from the earliest stages of project planning into the environmental review process.

- **Enhance legal sufficiency reviews**—uses the process to identify the most common problems in NEPA and Section 4(f) document development, their root causes, and the measures local and state transportation agencies can take to avoid the problems; and encourages reviews when documents are in their draft stage, reducing the potential need for multiple legal reviews of a "final" document and helping to resolve conflict and potential controversy earlier in the process, when project schedules can better accommodate the change.
- **Expand the use of programmatic agreements**—identify new and existing programmatic agreements that may be expanded to a regional or national level.
- **Encourage the use of existing regulatory flexibility**—clarify existing requirements applicable to activities that may be allowed during the preliminary design phase of development and to ROW acquisition and utility relocation.

These issues identified by both DOT and state transportation agencies illustrate the need to more efficiently implement existing requirements or to identify barriers to implementing them.

CONCLUSION

There is little debate that delays in transportation project delivery can result in higher project costs, as well as delay potential positive economic advantages such as bringing project-related jobs to the community. Also, it is known that completing the environmental review process takes time, sometimes years for complex, major projects. Meeting environmental compliance requirements may result in project delays or, at least, a project taking longer than anticipated by its sponsor. However, what is unclear is whether or what specific elements of the environmental review process routinely delay project delivery.

The time it takes to complete the NEPA process is often the focus of debate over project delays attributable to the environmental review process. However, the influence of environmental requirements established under Title 23 and other federal law call into question the degree to which changes in the NEPA process will expedite the environmental reviews and accelerate project delivery. Further, although there are no comprehensive data and available information tends to be anecdotal, when delays in the environmental review process have been identified, they primarily stemmed from local or project-specific issues (e.g., project complexity, changes in state priorities, or late changes in project scope).

Regardless of potential changes to the NEPA process or the overall environmental review process, local factors will strongly influence project delivery time. State or local decision makers will continue to have the most significant influence on project delivery in their capacity to establish (and change) project priorities, allocate available funds, and be influenced by local controversy or project opposition. A project's environmental review process may be efficiently executed and involve no delays in the process itself, but still take decades or never be completed if local and state issues are acting against the project.

The potential success of efforts intended to expedite the environmental review process would involve evidence that transportation projects were delivered more quickly. However, considering the limits to measuring the time it takes to complete the environmental review process, the relative success of a particular approach may be gauged in terms of the degree to which state or local transportation agencies find it useful in meeting their environmental compliance obligations.

Compared to transportation planning and project development during construction of the Interstate Highway System, state and local transportation agencies are more inclined to consider a project's effects on communities and resources. Apart from any potential changes to federal environmental review requirements, local and state agency decisions regarding transportation project planning, funding, and development will continue to be strongly influenced by a project's benefits and adverse effects to the environment and the community it serves.

APPENDIX. SURVEYS AND STUDIES APPLICABLE TO THE ENVIRONMENTAL REVIEW PROCESS

In this report, summary information and conclusions regarding factors applicable to measuring the stages of project development, the time it takes to complete the environmental review process, and primary sources of delay or perceptions among transportation agency officials regarding causes of delay in completing the environmental review process were drawn from data included in the following surveys and studies conducted by FHWA, GAO, universities, or transportation organizations:

Federal Highway Administration (available on FHWA's "Environmental Toolkit: Streamlining/Stewardship—Performance Reporting" website, http://environment.fhwa.dot.gov/strmlng/es10measures.asp).

- *Evaluating the Performance of Environmental Streamlining:* Phase II, an FHWA-commission study conducted by the Louis Berger Group, 2003.
- FHWA surveys, *Reasons for EIS Project Delays and Information on Timeliness on Completing the NEPA Process.*
- *Strategies and Approaches for Effectively Moving Complex Environmental Documents Through the EIS Process:* A Peer Exchange Report, prepared for FHWA by DOT's John A. Volpe National Transportation Systems Center Research and Innovative Technology Administration, January 2009.
- *FHWA/Gallup Study on Implementing Performance Measurement in Environmental Streamlining,* "Implementing Performance Measurement in Environmental Streamlining," May 2007.

Government Accountability Office.

- *Highway Infrastructure: Stakeholders' Views on Time to Conduct Environmental Reviews of Highway Projects*, GAO-03-534, May 23, 2003.
- *Highway Infrastructure: Preliminary Information on the Timely Completion of Highway Construction Projects*, GAO-02-1067T, September 19, 2002.

University and Transportation Organization Studies.

- *What Influences the Length of Time to Complete NEPA Reviews? An Examination of Highway Projects in Oregon and the Potential for Streamlining*, by Jennifer Dill, Center for Urban Studies, Nohad A. Toulan School of Urban Studies &Planning, Portland State University, submitted for presentation at the 85th Annual Meeting of the Transportation Research Board, November 15, 2005 (revised).

- *Causes and Extent of Environmental Delays in Transportation Projects*, prepared by TransTech Management, Inc., for the American Association of State Highway and Transportation Officials (AASHTO), December 2003.
- Environmental Streamlining: A Report on Delays Associated with the Categorical Exclusion and Environmental Assessment Processes, prepared by TransTech Management, Inc., for AASHTO, October 2000.

End Notes

[1] This report focuses on projects approved under programs administered by FHWA. Although they involve similar regulatory requirements, issues unique to transit projects approved under programs administered by the Federal Transit Administration (FTA) are not addressed in this report.

[2] Those requirements are largely established under Chapter 1, "Federal-aid Highways," of Title 23, "Highways" of the *U.S. Code.*

[3] In this report, reference to "federal-aid highways," "federal highways," or "federal highway projects" means projects that may receive federal funding pursuant to the Federal-aid Highways provisions of Title 23. Those projects include, but are not limited to, the initial construction, reconstruction, replacement, rehabilitation, restoration, or other improvements of a highway, road, street, parkway, right-of-way, bridge, or tunnel.

[4] See CRS Report R41947, *Accelerating Highway and Transit Project Delivery: Issues and Options for Congress*, by William J. Mallett and Linda Luther.

[5] The Safe, Accountable, Flexible, Efficient Transportation Equity Act: A Legacy for Users (SAFETEA-LU or SAFETEA; P.L. 109-59, for FY2005-FY2009) and the Transportation Equity Act for the 21st Century (TEA-21; P.L.105-178, for FY1998-FY2003).

[6] H.R. 7 was reported favorably by the House Transportation and Infrastructure Committee on February 13, 2012. MAP-21 passed the Senate on March 14, 2012.

[7] See *NEPA and Project Development: Program Overview* on FHWA's "Environmental Review Toolkit" webpage at http://environment.fhwa.dot.gov/projdev/index.asp.

[8] 23 C.F.R. Part 771.

[9] The National Highway System is approximately 160,000 miles of roadway important to the nation's economy, defense, and mobility.

[10] For more information about the Interstate Highway System, see the U.S. Department of Transportation's Federal Highway Administration website, "Celebrating the Eisenhower Interstate Highway System," http://www.fhwa.dot.gov/interstate/homepage.cfm.

[11] Report prepared by the John A. Volpe National Transportation Systems Center Research and Innovative Technology Administration, U.S. Department of Transportation for the Office of Project Development and Environmental Review, Federal Highway Administration, "Strategies and Approaches for Effectively Moving Complex Environmental Documents Through the EIS Process: A Peer Exchange Report," January 2009, available on the Federal Highway Administration's "Streamlining/Stewardship" website at http://environment.fhwa.dot.gov/strmlng/eisdocs.asp.

[12] For a discussion of issues related to freeway revolts and general issues with urban freeway construction, see "Paved with Good Intentions: Fiscal Politics, Freeways and the 20th Century American City," by Jeffrey A. Brown, Eric A. Morris, and Brian D. Taylor, in the

University of California Transportation Center's *Access* magazine, Fall 2009, available at http://www.uctc.net/access/35/access35_Paved_with_Good_Intentions_Fiscal_Politics_.shtml.

[13] The term "Section 4(f)" refers to the section of the Department of Transportation Act of 1966 (P.L. 89-670) under which the requirement was originally set forth. It was initially codified at 49 U.S.C. §1653(f) and only applied to DOT agencies. Later that year, 23 U.S.C. §138 was added with somewhat different language, which applied only to the highway program. In 1983, as part of a general recodification of the DOT Act, §1653(f) was formally repealed and codified in 49 U.S.C. §303 with slightly different language. This provision no longer falls under a "Section 4(f)," but DOT has continued this reference, given that over the years, the whole body of provisions, policies, and case law has been collectively referenced as Section 4(f).

[14] Depending on project alternatives under consideration for a given project, compliance with Section 4(f) requirements can be complex. This report does not discuss those requirements in detail. For more information, see the "Section 4(f)"website included in FHWA's "Environmental Review Toolkit," at http://www.environment.fhwa.dot.gov/4f/index.asp.

[15] *Citizens to Preserve Overton Park v. Volpe*, 401 U.S. 402 (1971).

[16] See Department of Transportation, "Parks, Recreation Areas, Wildlife and Waterfowl Refuges, and Historic Sites," final rule, 73 *Federal Register* 13367-13401, March 12, 2008.

[17] Air quality issues are also relevant to federal-aid highway project development. Under the Clean Air Act, FHWA must insure that transportation plans, programs, and projects conform to the state's air quality implementation plans. Conformance with a state implementation plan is largely determined during project planning. Issues associated with meeting federal air quality requirements are not discussed in this report.

[18] 40 C.F.R. §§1500-1508.

[19] 40 C.F.R. §1507.3.

[20] 23 C.F.R. Part 771.

[21] These statistics apply to projects funded under the Federal-aid Highway program. For more detail, see the Federal Highway Administration's "Highway Statistics for 2009: Obligation of Federal-Aid Highway Funds For Highway Improvements Fiscal Year 2009 (Intended to Show Only Projects Authorized in FY 2009)," Table FA-10, October 2010, available at http://www.fhwa.dot.gov/policyinformation/statistics/2009/fa10.cfm..

[22] That definition of "Major Projects" was included among provisions in Section 1904 of SAFETEA that amended the "Project approval and oversight" requirements under 23 U.S.C. §106. The identification of a project as "major," in this context, is unrelated to its potential distinction as a "major federal action significantly affecting the quality of the human environment" pursuant to NEPA under 42 U.S.C. §4332(c).

[23] See the *FHWA Active Project Status Report*, available on FHWA's "Project Delivery" website, https://fhwaapps.fhwa.dot.gov/foisp/publicActive.do.

[24] 23 C.F.R. §771.117(a); further DOT criteria used to determine whether a project would meet necessary CE criteria extend from CEQ regulations defining CEs at 40 C.F.R. §1508.4.

[25] 23 C.F.R. §771.117(b).

[26] Listed at 23 C.F.R. §771.117(c).

[27] Listed at 23 C.F.R. §771.117(d).

[28] 23 U.S.C. §326.

[29] For more information, see memorandum from the U.S. Department of Transportation, Federal Highway Administration, to Directors of Field Services and Division Administrators, regarding "Guidance on the State Assumption of Responsibility for Categorical Exclusions (CE)," April 6, 2006, available at http://www.fhwa.dot.gov/hep/6004memo.htm.

[30] 23 C.F.R. §771.115(a).

[31] 40 C.F.R. §1508.22.

[32] 40 C.F.R. §1502.9.

[33] This category of agency participant in the NEPA process differs from a "cooperating agency," discussed below, that is defined as an agency having jurisdiction by law or special expertise with respect to any environmental impact of a proposed project or project alternative.

[34] See "SAFETEA-LU Environmental Review Process, Final Guidance," November 15, 2006, available at http://www.fhwa.dot.gov/hep/section6002/index.htm and the Department of Transportation's "Environmental Impact and Related Procedures; Final Rule," 74 *Federal Register* 12517, March 24, 2009.

[35] See the National Cooperative Highway Research Program's "Legal Research Digest 54: Practice Under the Environmental Provisions of SAFETEA-LU," December 2010, available at http://onlinepubs.trb.org/onlinepubs/ nchrp/nchrp_lrd_54.pdf.

[36] There was also wide approval of changes made to Section 4(f) under SAFETEA Section 6002, applicable to *de minimis* project impacts (see "Requirements Applicable to Federal-aid Highways" regarding the "preservation of parklands" requirements.

[37] See the summary of survey respondent impressions of SAFETEA provisions at pp. 16-21.

[38] 40 C.F.R. §1508.9.

[39] 23 C.F.R. §771.119(b).

[40] See 40 C.F.R. §1508.16.

[41] 23 C.F.R. §771.125(b). A legal sufficiency review of an EA may be required if FHWA determines that details of the individual project warrant such a review.

[42] 23 C.F.R. §774.7(d).

[43] 42 U.S.C. §4332(2)(C).

[44] 40 C.F.R. §1508.5.

[45] Specific only to DOT's NEPA requirements, "participating" agencies for federal highway projects are defined at 23 C.F.R. §771.107(h) as a state, local, tribal, or federal agency that may have an interest in the proposed project and have accepted an invitation to participate in the environmental review process.

[46] 40 C.F.R. §1503.1.

[47] 40 C.F.R. §1503.

[48] 40 C.F.R. §1503.4.

[49] Text of Secretary Mineta's May 6, 2003, letter, and Chairman Connaughton's May 12, 2003, response, are available at http://www.environment.fhwa.dot.gov/guidebook/Ginterim.asp.

[50] 40 C.F.R. §§1500.2(d). 1506.6.

[51] 23 U.S.C. §128.

[52] Plaintiffs have generally cited some inadequacy in the NEPA documentation as the basis for filing NEPA-related lawsuits (see CEQ's *Litigation Surveys* for each year from 2001 to 2009 on its "NEPA Litigation" web page at http://ceq.hss.doe.gov/legal_corner/litigation.html). They may charge, among other things, that an EIS or EA did not include sufficient analysis of all project alternatives, did not consider all "reasonable" project alternatives, did not adequately analyze the effects of project alternatives, or that an EA was prepared when an EIS should have been (i.e., a FONSI was issued when impacts were in fact significant).

[53] From 2001 to 2009, NEPA-related lawsuits filed annually against FHWA ranged from a low of three to a high of 12; see CEQ's *Litigation Surveys* cited in footnote 52.

[54] U.S. General Accounting Office (now the Government Accountability Office), *Highway Infrastructure: Preliminary Information on the Timely Completion of Highway Construction Projects*, September 19, 2002.

[55] In addition to information from FHWA and state DOT staff, this report also looked at the time it took and the steps necessary to complete six new highway construction projects in California, Florida, and Texas (the largest in the state, in terms of federal funds received, and a randomly selected "medium-sized" project).

[56] Consistent with the factors that make it difficult to measure individual phases of project development, discussed above, GAO noted that federal and state governments do not maintain information centrally (or, in some cases, at all) on the time it takes to complete highway projects. GAO also noted that there was no accepted measuring stick with which to

gauge whether project performance is "timely." To make its determination on project timing, GAO relied on a best estimate prepared by FHWA. According to FHWA, the estimate it provided to GAO was based on the professional judgment of its staff and several state DOTs.

[57] The identification of factors that currently affect project delivery is particularly relevant when considering legislative options to address potential causes of delay in the environmental review process. State DOTs have improved their environmental review procedures in the past 10 years. Also, FHWA has expanded its efforts to provide information and guidance on the process, including increased efforts to encourage states to implement programmatic agreements applicable to NEPA compliance and other environmental laws.

[58] Federal Highway Administration and the Louis Berger Group, *Evaluating the Performance of Environmental Streamlining: Phase II*, 2003, available on FHWA's "Streamlining/ Stewardship" website, http://www.environment.fhwa.dot.gov/strmlng/baseline/phase2rpt. asp.

[59] Ibid., under "Conclusions 4.2."

[60] Factors listed here are those that have been most commonly identified in surveys or studies conducted by FHWA and GAO, as well as selected university and transportation organizations. For a list of the surveys and studies used to prepare this report, see **Appendix**. Those surveys and studies have looked primarily at causes of delay applicable to projects that require an EIS.

[61] See "FHWA/Gallup Study on Implementing Performance Measurement in Environmental Streamlining," available at http://environment.fhwa.dot.gov/strmlng/gallup_05-07.asp.

[62] Thomas, H.R. and Ellis, R.D, *Avoiding Delays During the Construction Phase of Highway Projects*, Transportation Research Board, National Research Council, October 2001, NCHRP 20-24(12). Also see "The Root Causes of Delays in Highway Construction," a summary of the study's findings submitted for presentation by the authors at the TRB annual meeting in 2003, available at http://www.ltrc.lsu.edu/TRB_82/TRB2003-000646.pdf.

[63] "The Root Causes of Delays in Highway Construction," p. 3.

[64] *Strategies and Approaches for Effectively Moving Complex Environmental Documents Through the EIS Process: A Peer Exchange Report*, prepared by the John A. Volpe National Transportation Systems Center Research and Innovative Technology Administration, U.S. Department of Transportation for the Office of Project Development and Environmental Review, Federal Highway Administration, January 2009, available on FHWA's "Streamlining/Stewardship" website at http://environment.fhwa.dot.gov/strmlng/eisdocs. asp.

[65] Ibid., under "Recommendations for Successful Tools & Techniques."

[66] See "Meeting Environmental Requirements After a Bridge Collapse," prepared for the Office of Project Development and Environmental Review, Federal Highway Administration, U.S. Department of Transportation, and prepared by the John A. Volpe National Transportation Systems Center Research and Innovative Technology Administration, U.S. Department of Transportation, August 2008, available at http://www.environment.fhwa.dot.gov/projdev/ bridge_casestudy.asp.

[67] Major sources of project delay cited in the bridge study are those identified in FHWA survey results included in **Appendix**.

[68] For more information about the bridge reconstruction project, see the Minnesota DOT "I-35W St. Anthony Falls Bridge" website at http://projects.dot.state.mn.us/35wbridge/index.html.

[69] See findings in the 2007 FHWA/Gallup study (cited in footnote 61).

[70] CEQ guidance "Improving the Process for Preparing Efficient and Timely Environmental Reviews under the National Environmental Policy Act," released on March 6, 2012, available at http://www.whitehouse.gov/administration/eop/ceq/initiatives/nepa/efficiencies-guidance. CEQ stated that the guidance is part of its broader effort to "modernize and reinvigorate" federal agency implementation of NEPA and to support goals established in

President Obama's August 31, 2011, memorandum, "Speeding Infrastructure Development through More Efficient and Effective Permitting and Environmental Review." For information about CEQ pilot programs established to support those goals, see http://www.whitehouse.gov/administration/eop/ceq/initiatives/nepa/nepa-pilot-project.

[71] See FHWA's online "Environmental Review Toolkit," available at http://environment.fhwa.dot.gov/index.asp. It includes, for example, guidance and information regarding linking project planning and environmental requirements; NEPA requirements applicable to project development; a database of "lessons learned" related to streamlining and environmental stewardship; and guidance on compliance requirements such as those applicable to wetlands, Section 4(f), and historic preservation.

INDEX

D

E

I

J

L

M

N

O

P

Q

R

S

T

U

V

W